PRACTICAL ORGANIC CHEMISTRY –A Primer

(QUALITATIVE ANALYSIS AND PREPARATION OF ORGANIC COMPOUNDS)

PRACTICAL ORGANIC CHEMISTRY –A Primer

(QUALITATIVE ANALYSIS AND PREPARATION OF ORGANIC COMPOUNDS)

Dr. Venkateswarlu Peesapati,
MSc, PhD, CChem, FRSC, FTAS
UGC Professor (Retd.) Osmania & J N T University,
& Visiting Prof. NIPER, Hyderabad, TS.

PharmaMed Press
An imprint of BSP Books Pvt. Ltd.
4-4-309/316, Giriraj Lane,
Sultan Bazar, Hyderabad - 500 095.

Practical Organic Chemistry – A Primer

by Dr. **Venkateswarlu Peesapati**

Published by

PharmaMed Press

An imprint of BSP Books Pvt. Ltd.

4-4-309/316, Giriraj Lane, Sultan Bazar, Hyderabad - 500 095.

Phone: 040-23445688; Fax: 91+40-23445611

E-mail: info@pharmamedpress.net

www.bspbooks.net/www.pharmamedpress.net

ISBN: 978-81-961469-6-2 (Hardback)

This text is gratefully dedicated to my wife

Mrs. P. V. Lakshmi

Foreword

This book is written in order to serve the needs of students who will be joining three years degree course after completion of 10+2 curriculum. These students do not have adequate exposure to practical organic chemistry. Keeping this in mind, the practical guide has been written. This book is designed to help the students to learn the basic skills involved in the identification of unknown organic compounds and also to show them how practical organic chemistry relates to theoretical aspects. To start with a student is made to know what different types of apparatus are used in the laboratory. Therefore a separate chapter was introduced giving diagrammatic representations of the different glass apparatus and their names. By reading the experimental procedure, a student can choose an appropriate glass ware needed for his/her experiment. All glass ware should be thoroughly cleaned and dried before the start of the experiment. Therefore the author, meticulously described, how to prepare cleaning mixture and how to dry the apparatus etc., in the beginning itself. Before taking any sample for qualitative or spectroscopic analysis, the product must be pure. Hence he has discussed the various methods of purification including chromatographic separation methods followed by various criteria to establish the purity of the sample.

In the next chapter the author described meticulously the reactions of various functional groups present in the carbon compounds in the form of a table. Now- a- days spectroscopic identification of organic compounds has become part and parcel of organic chemistry syllabus. Therefore, after tentatively identifying the functional group by chemical tests, some diagnostic spectral features (uv, ir, n.m.r, &^{13}C n.m.r) were given in the form of a table. Thus it becomes easy to confirm the presence of particular functional group with spectral data. The student will be able to confirm the identity of the unknown compound by preparing rational derivatives.

A part of the book is devoted to the preparation of simple and essential compounds by well known name reactions and where ever it is necessary the mechanism involved in the preparation of these compounds are discussed. Majority of these experiments are class tested. The author included the physical constants such as mp, bp, of various derivatives. A small section of this book describes the preparation of very commonly encountered reagents. At the end of the book some problems and solutions are included.

As explained in this book, if the students undergo training we believe that they will be able to understand, better the relationship between theory and practicals in organic chemistry. This book will definitely help the students to gain experience at the initial level of UG and give them an advantage when they enter into higher level in their career in chemistry.

Lastly, the book not only helps the students but also give sufficient training for fresh teachers, who are entering into chemistry teaching.

Prof. P N Sarma
Retd. Professor Osmania University,
Hyderabad, TS.

Preface

There is no better device than classical qualitative organic analysis for developing a real understanding and appreciation of organic chemistry. The purpose of the beginning undergraduate organic laboratory curriculum is to provide students with practical skills in studying preparative observational and interpretation experiences. The purpose is to provide them with the wide range of experience possible and to inculcate an appreciation of the largest number of substances, reactions, and techniques in the shortest possible time and minimum expenditure of labour and material. The essence of this is simplicity.

Students who will be joining three years degree course (after completing two years intermediate) with an intention to learn Organic Chemistry do not have adequate exposure to practical organic chemistry. Keeping this in mind of such students, this practical guide has been written. This book contains five chapters.

The 1st chapter describes various experimental methods followed in organic chemistry. In this chapter, the required apparatus in the laboratory, characteristics of pure carbon compounds and their determination by different methods used for purification and related topics are discussed in detail. Most modern techniques like thin layer, paper, and column chromatography methods are described in such a way, that the students will be able to carry out these techniques by themselves.

The chapter 2 deals with chromatographic separation of mixtures using different methods, t.l.c, paper and column chromatography.

Reaction of carbon compounds containing different functional groups are discussed in chapter 3.

3rd chapter also deals with qualitative organic analysis which forms the main topic. Preliminary test of carbon compounds, classification of compounds based on solubility tests followed by group tests are mainly discussed in this chapter.

Only after performing thoroughly the tests for various functional groups by taking known organic compounds, the students should go for qualitative organic analysis of unknown compounds.

To confirm their identification, students must prepare appropriate derivatives. The mps & bps of important simple organic compounds and their derivatives are given in section 3.7

The 4th chapter describes the preparation of common and easily available organic compounds. Teachers must advice the students about the preparations already made as starting materials to make other compounds to the extent possible.

In 5th chapter the basic concepts of spectroscopy such as ultra violet (UV), infra-red (IR) spectroscopy, nuclear magnetic resonance (NMR) and mass spectrometry are discussed with few examples for each category.

In addition to these five chapters, eight annexures are added at the end of the experimental book. Making different types of solutions, buffer solutions how to effect dilutions, to make dilute acids & bases are described in annexure I.

Some important laboratory reagents and chemicals used in the preparation of derivatives are included in annexure II & III respectively. Risk management in the usage of organic solvents and inorganic reagents is described in annexure IV.

Annexure V describes the safety measures to be adopted if accidents occur in the laboratory. What items of the apparatus must be issued to each student are listed in annexure VI and Annexure VII Details of Elements are provided.

Lastly study questions with solutions have been included in Annexure VIII.

We trust, that students will be able to understand organic chemistry easily, if they are trained in experimental methods, as explained in this book.

Dr. Venkateswarlu Peesapati

Hyderabad (TS)

Acknowledgements

The author is extremely grateful to Dr. P. N. Sarma, Retd. Professor, OU, for meticulously going through the entire draft of the manuscript and enriching it with his valuable comments and editing, which brings completeness to the book. I thank him for readily providing forward to the book.

Thanks to Dr. B. Rupavani, Dr. K. Anuradha and Dr. N. Shankaraiah for their help in drawing structures and Mrs. T. Neelima for her help in typing the manuscript.

My special thanks to my former colleague Dr. Norrie Wilson, Edinburgh University, Scotland for his valuable comments during the process of writing the book. Thanks to Dr. Y. V. D. Nageswar for his valuable suggestions.

Thanks to Mr. Venkateswar Rao, K. V. Xerox for attending to every request of mine to print and binding the manuscript during the preparation of the book.

My sincere thanks to Mr. Anil Shah, Director PharmaMed Press and Mr. Naresh Davergave, Production Manager & other staff for their support in bringing this book.

The author wishes to place on record his grateful thanks to his wife Mrs. P. V. Lakshmi for involving and organizing the script in all stages. Without her help this work would not have been possible.

Thanks to my sons, Mr. Sreedhar and Mr. Vidyadhar & daughter-in-laws, Mrs. Caroline and Mrs Carly for their encouragement.

Above all, to my beloved teacher and PhD guru, Padmabhushan, late Prof. T. R. Seshadri, FRS, for stimulating a lifelong love of chemistry.

Acknowledgements

CONTENTS

CHAPTER 1

Experimental Methods and Manipulation

1.1 Introduction

Introduction: This book will help the students to learn the basic skills involved in the identification of unknown organic compounds and show them how practical chemistry relates to theoretical work. The experiments are designed to be straight forward, and have been well tested for reproducibility and reliability.

The following procedure will ensure a student's success in the laboratory.

Always know exactly what you are trying to do!

(i) Read all the details of the experiments before the laboratory class, then if you have any doubts you can ask the demonstrator before you make a mistake.

(ii) Carry out the identification and structure assignment of the given unknown compound.

(iii) Write up each experiment or analysis as soon as it is finished.

(iv) Ensure that you are able to answer in the laboratory note-book any questions asked at the end of the experiment. These questions are designed to test your understanding of the experimental procedures and theory behind it.

(v) Treat your samples with care. Label properly all the specimens of derivatives that you have made.

Do not rush for anything, and if you have any doubt ask demonstrator.

1.2 Apparatus used in the Laboratory

Every student should own a pair of safety glasses or goggles and should be required to wear these at all times in the laboratory. Accidents involving the eyes are extremely serious and usually arise most unexpectedly and from trivial causes.

APPARATUS

Glass Apparatus:

Tripod stand

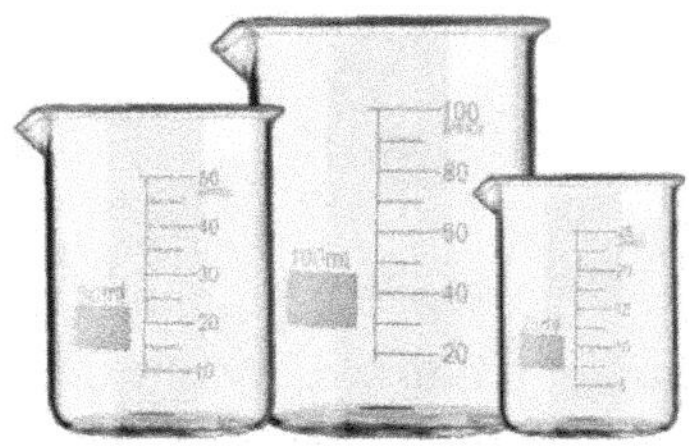

Graduated beakers

Spirit lamp

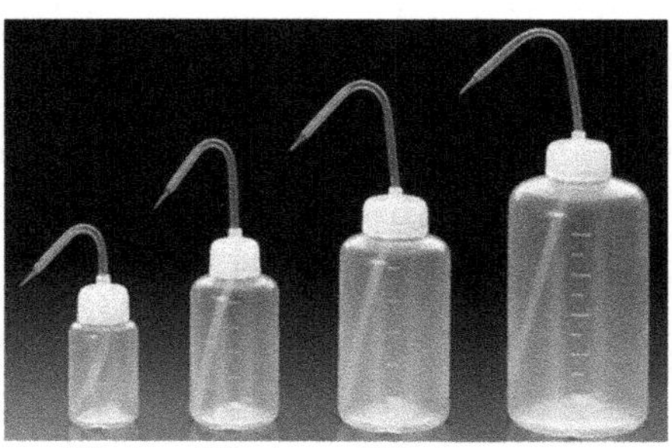

100ml 250ml 500ml 1l

Polythene Wash bottles

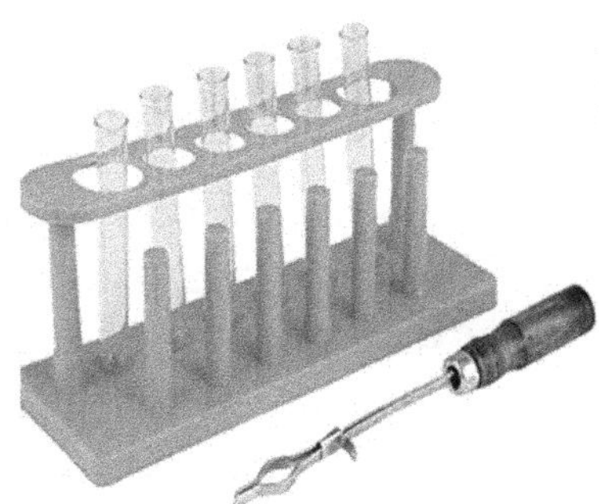

Test tubes stand with holder

Forceps

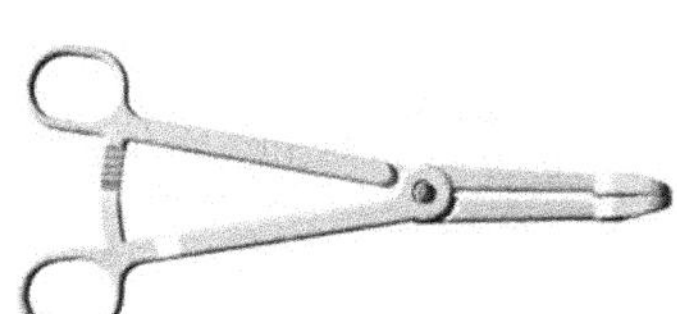

Tongs

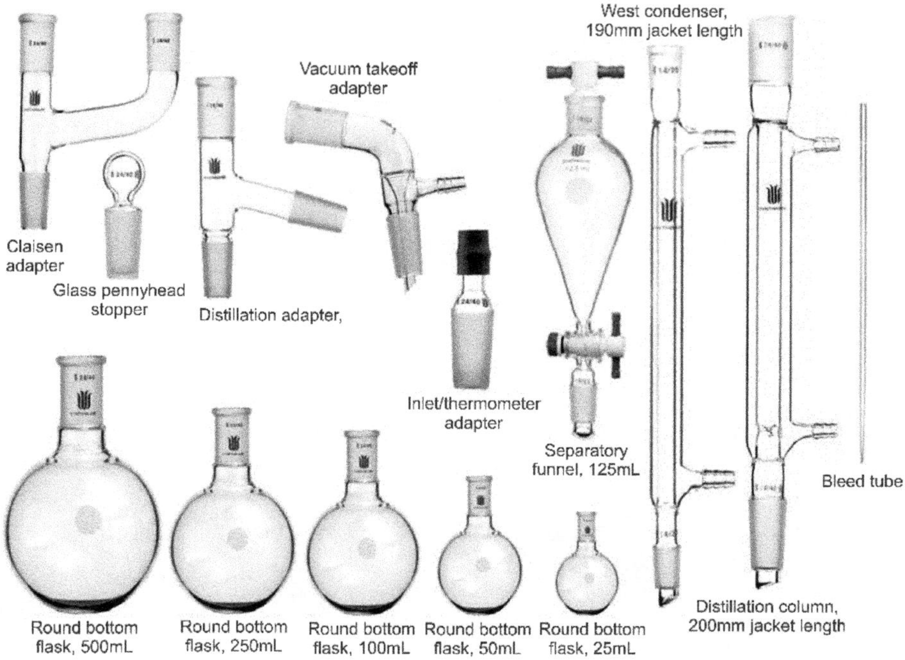

Watch Glass

Spatula

Wire Guaze

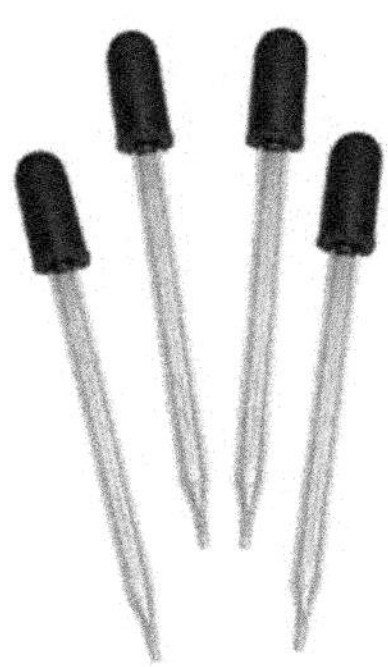
Droppers

Boiling Tubes With Stand

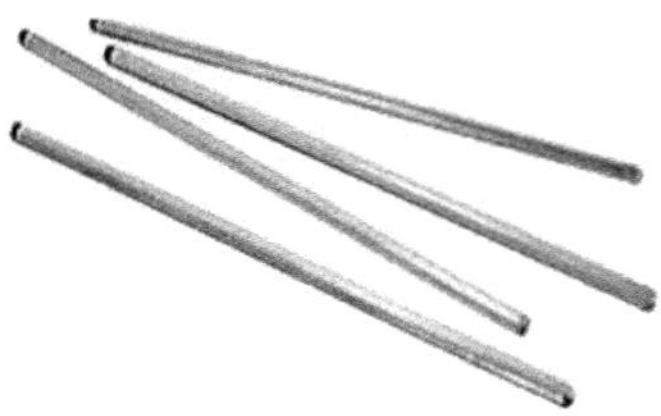

Glass Rods

Vacuum Desiccator

Magnetic Stirrer with Hot Plate

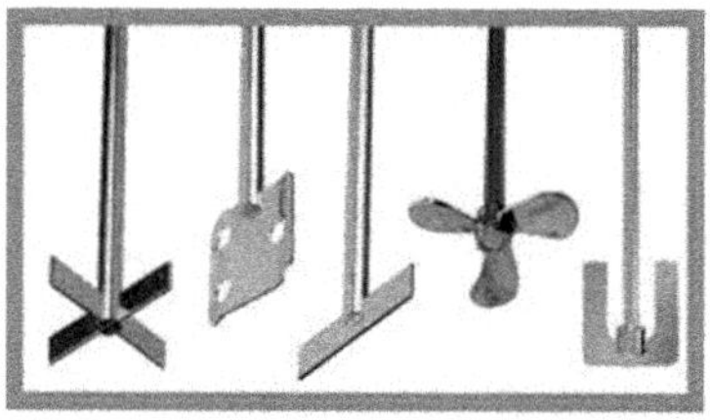

Different Types of Stirrers

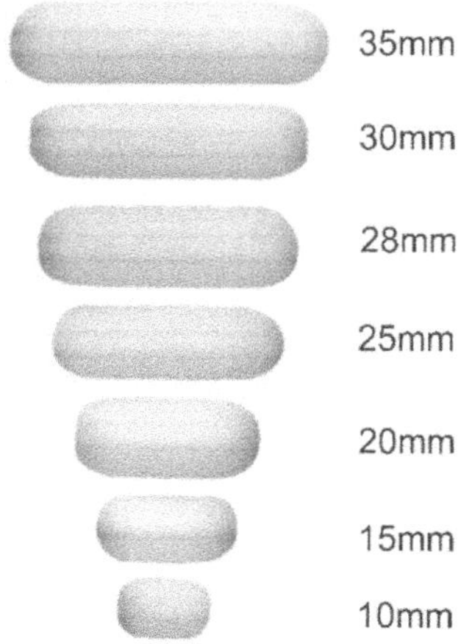

Teflon Magnetic Bars

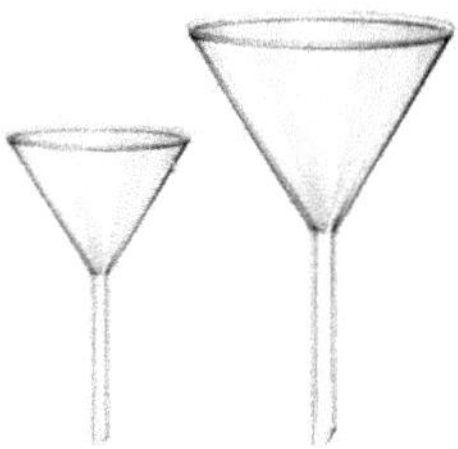

Glass Funnels

Filtration Set

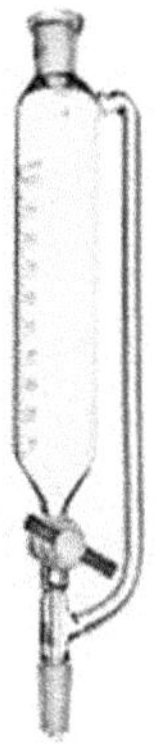

Dropping Funnel

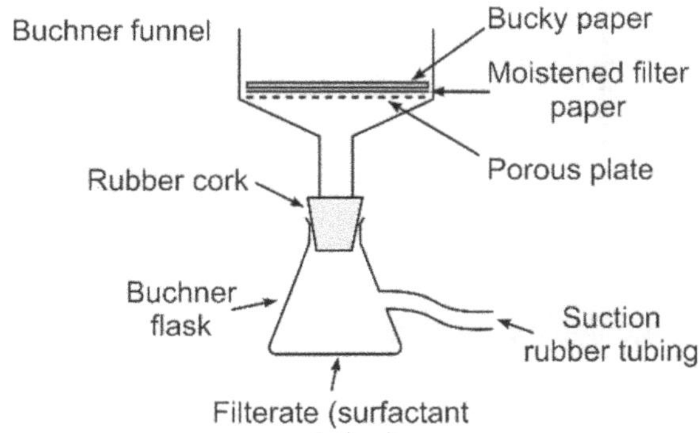

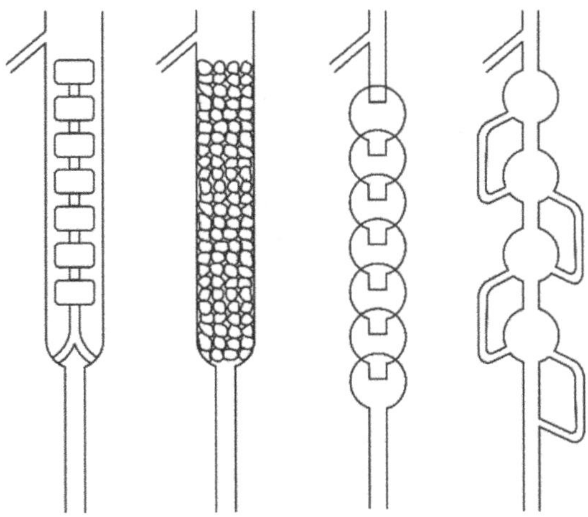

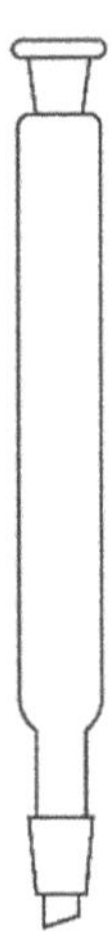

open column

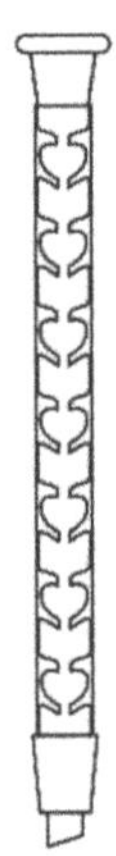

Vigreux column

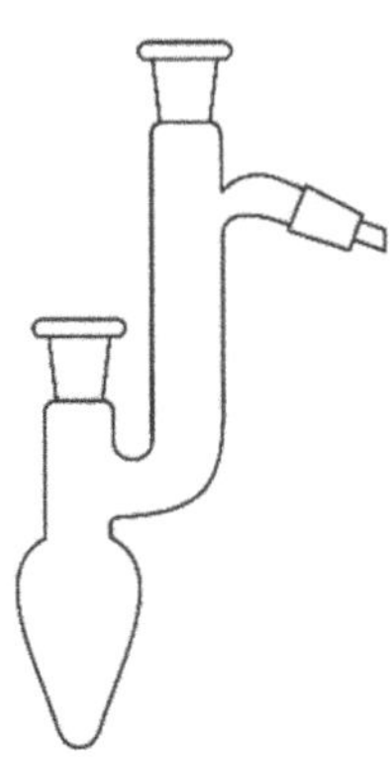

Claisen flask

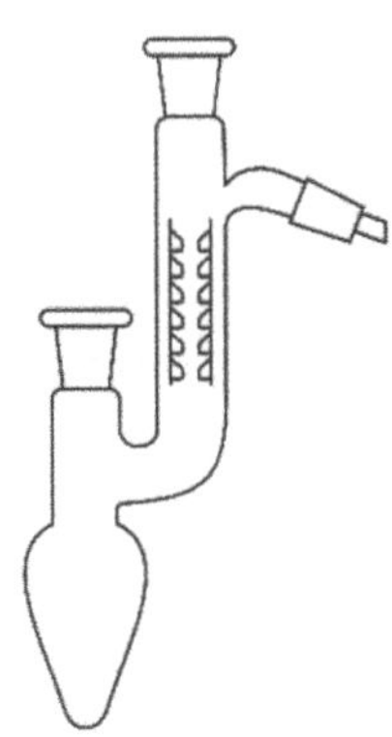

Claisen-Vigreux flask

Retort stand

Iron rings

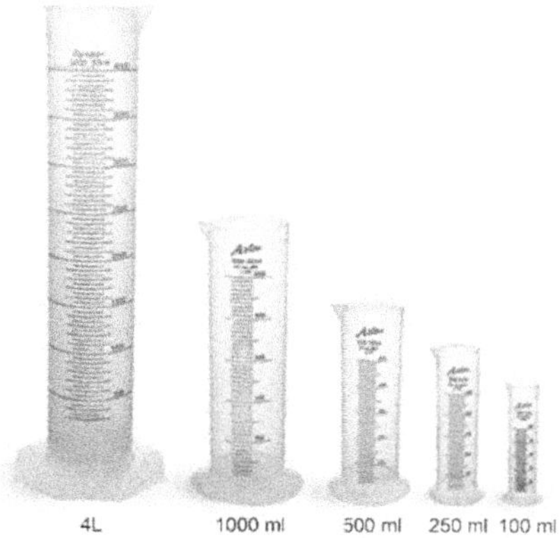

Graduated measuring jars

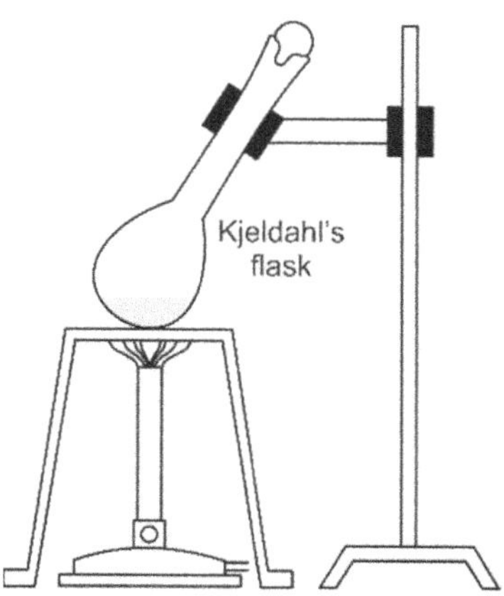

Kjeldahl flask (mp)

Graduated pipette

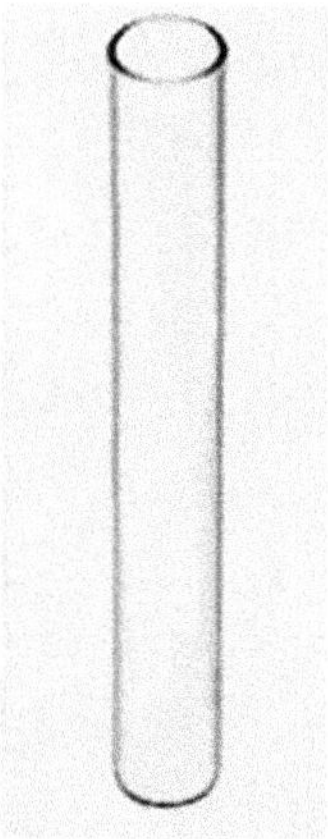

Boiling Tube

Sample tubes

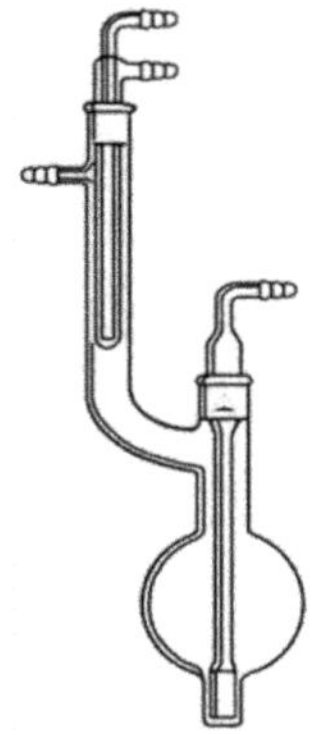

Hydrogenation
Apparatus (Micro scale)

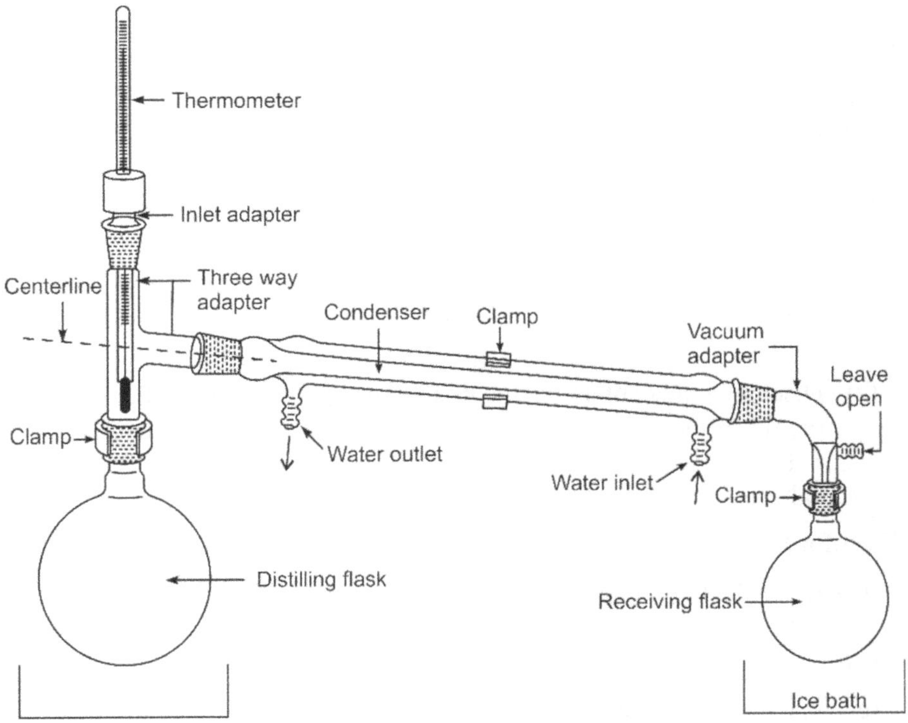

Distillation apparatus

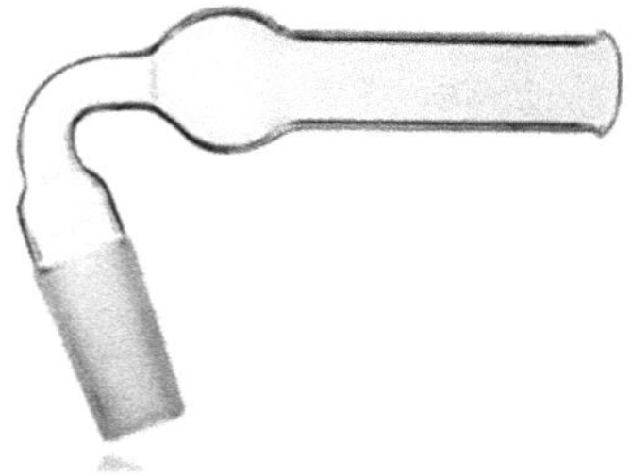

Guard Tube (to be filled with dry $CaCl_2$)

Melting point apparatus

1.3 Cleaning Glass Apparatus

Use of Pipette

(a) Use clean glassware at all times

(b) Before pipetting a solution, rinse the pipette using 2-5 ml of the solution to be measured. Discard the solution.

(c) Using a pipette bulb fill the pipette with the solution so that it is just above the calibration mark.

(d) Press the valve on the pipette with the solution to run down so that the shadow of the bottom of the meniscus is level with the calibration mark.

(e) Discharge the contents of the pipette into the appropriate flask. Touch the tip of the pipette against the surface of the liquid in the flask. Do not blow out the last drop of liquid from the pipette, the pipette is calibrated to account for this drop!

Use of a Burette

(a) Rinse the burette using 2-5 ml of the solution to be measured. Discard the solution.

(b) Have a wash bottle with distilled water ready for washings.

(c) Place a small funnel in the top of the burette and fill with the solution.

(d) Remove the funnel

(e) Open the burette tap so that the portion below the tap is full of solution to be used with no air bubbles.

(f) Make a "sail" with a filter paper to give a white back ground against which the meniscus can be viewed. Place a filter paper under the flask containing the solution to be analyzed to give a white back ground to view color changes.

(g) It should be able to record the level of the burette to two decimal places.

(h) Add the solution to the reaction flask. Regularly swirl the flask to ensure good mixing.

(i) Rinse down the sides of the flask with distilled water during the titration to ensure all of the reactants are utilized.

(j) As the end point is approached decrease the rate of addition. You should be able to add half a drop by touching the tip of the burette with the reaction flask and rinse down the sides of the flasks. Swirl the flask after each addition. Ensure there are no bubbles in the tip.

(k) Carry out all titrations in duplicate to check for consistency. The titre values should be consistent ± 0.1.

(l) All glassware should be thoroughly cleaned and dried before the experiment. It is advisable to develop the habit of cleaning all glass apparatus immediately after their use. Otherwise, it becomes difficult if the dirty apparatus is allowed to stand for a longer period, particularly if volatile solvents have evaporated in the mean time.

The simplest method for gross deposits, when access to a test-tube brush is possible, is to employ a commercial household washing powder containing an abrasive which does not scratch glass (e.g. Vim, Dish-wash, etc.). Finally, the apparatus is thoroughly rinsed with water and dried in an air oven.

Alternatively the apparatus can be cleaned by using the 'chromic acid' cleaning mixture. This is essentially a mixture of chromium trioxide and concentrated sulfuric acid, and possesses powerful oxidizing and cleaning properties.

Preparation of chromic acid solution: Five grams of sodium dichromate are wetted with 5 ml of water in a 250 ml beaker, 100 ml of concentrated sulfuric acid are then added slowly with constant stirring. The temperature will raise to 70-80 ^{0}C. The mixture is allowed to cool to room temperature and then transferred to a dry, glass-stoppered, clean labeled bottle. Before using this mixture for cleaning purpose, the flask to be cleaned should be rinsed with water to remove water soluble organic matter. After draining away as much of the water as possible, a small quantity of the cleaning mixture is introduced into the flask, the soiled surface thoroughly wetted with the mixture. After standing for a short period with occasional rotation of the flask to spread the liquid over the surface, the flask is thoroughly washed with tap water and finally rinsed with distilled water.

1.4 Drying of Organic Solutions (Extracts)

The combined solvent phases resulting from extraction of an aqueous solution will contain small amounts of water. This is usually removed before the solution is evaporated, a process usually accomplished by shaking the solution with anhydrous drying agent, which will absorb the water but do not react with the solvent or the desired solute.

***Five suitable drying agents are listed below*:**

(i) Anhydrous magnesium sulphate ($MgSO_4$)

(ii) Anhydrous sodium sulphate (Na_2SO_4)

(iii) Anhydrite calcium sulphate ($CaSO_4$)

(iv) Anhydrous potassium carbonate (K_2CO_3)

(v) Anhydrous calcium chloride ($CaCl_2$)

It is often convenient to remove final traces of water with the aid of molecular sieve and store the dried solvent in the presence of the sieve.

The term molecular sieve, applies to a group of dehydrated synthetic sodium and calcium aluminosilicate adsorbents; three types of Zeolites namely 3A, 4A, 5A and 13A are available commercially.

To dehydrate the solvent phase, add to it one-twentieth of its bulk of the selected desiccant and shake the mixture frequently during 10-20 minutes ($MgSO_4$) or allow it to stand with occasional shaking for 30 minutes or more (Na_2SO_4). Separate the dried solution from the spent desiccant either by filtration with suction or by means of a filter paper funnel or by careful decantation.

1.5 Solid Substances-Criteria for Purity

(A) Melting points: Calibration of a thermometer - Generally while determining the melting points, boiling points and reaction temperatures a thermometer is invariably used in the laboratory. In some situation, if the temperature is approximately determined, it does not matter. But sometimes we may have to determine the temperatures accurately. By using ordinary laboratory thermometer, we can determine the temperatures from -38 ^{0}C to 360 ^{0}C. Nearly all the boiling points and melting points of organic compounds fall in this range. Thermometers employed in the laboratory differ from one to other while indicating temperatures. If a given thermometer shows an error of 0.5 ^{0}C in the range of 0^0 to 100 ^{0}C, and an error of 1^{0}C in the range of 100 to 250 ^{0}C, we can accept it. But in some thermometers the value of temperatures determined, may fall exactly in the expected region, while the temperature values in higher temperatures regions may show 3-4 ^{0}C error. Because of these reasons, the student must calibrate, the thermometer given to him/her. The b.ps and m.ps of some pure compounds are taken as standard reference for temperatures. By determining the m.ps/ b.ps of such pure compounds with a given thermometer, we can find out the error in that particular thermometer.

The following compounds are taken as standard reference compounds and their mps & bps are given in the following table.

S.No.	Compound	Mp (^{0}C)	Bp (^{0}C)
1	Ice	0^0 C	----
2	Water	----	100^0C
3	Diphenylamine	53.5	----
4	Benzene	----	80.2
5	Vanillin	82	----
6	Resorcinol	112	----
7	Benzoic acid	122.4	----
8	Chlorobenzene	----	132
9	Salicylic acid	158.3	----
10	Hydroquinone	170	----
11	Succinic acid	189	----
12	3,5-Dinitro benzoic acid	205	---
13	Nitrobenzene	---	210.9
14	P-Nitrobenzoic acid	239	----

Using a given thermometer, determine the Mps/Bps of the above compounds & find out the error and record data in the following table:

Example:

S.No.	Standard compound used for calibration	Mp/Bp of pure compound	Temperature shown by thermometer	Error (in degrees)
1.	Hydroquinone	170^0	174^0 say	$+4^0$

(B) Determination of melting points: *The melting point is the prop*erty of an organic solid which is most frequently used as a criterion of purity. A pure compound has a sharp melting point (i.e. melts over + or – 1^0C temperature range).

Procedure: Crush a small amount of the sample on a watch glass with a small spatula. Cut double-sealed melting point capillary in to half or use commercially available melting point capillaries. Introduce a little of the powder into the open end of a capillary tube and tap it down to the closed end by gently stroking with a file. The column of solid should be no more than 3 mm in length and should be tightly packed. Select a thermometer with a short bulb (0-360^{0}C). Do not rely on digital displays. Record the temperature at which melting begins, and the temperature at which the last traces of solid disappear. This is the melting point range of the sample. You will not obtain accurate melting points if the sample is heated too rapidly.

Here one can use Kjeldahl flask for determining the melting points in addition to commercial melting point apparatus.

(C) **Mixed melting points:** Pure organic compounds exhibit sharp melting points. If they are contaminated with impurities, their melting points will be lower when compared to the melting point of pure compounds.

1.6 Liquid Substances - Determination of Boiling Point on a Semi-micro Scale

Boiling points of pure liquids

Procedure Part 1

Using a small Bunsen flame, draw out a centre part of a melting point capillary tube so that its diameter is reduced to about one-third of the original. To do this, hold the capillary tube horizontal between your hands gripping the tube at either end with a thumb or forefinger. While rotating the tube backwards and forwards, heat the centre section in the side of a blue flame, about 2 cm up from the flame outlet. The tube will glow orange and suddenly soften. Immediately pull out of the flame and move your hands apart in one continuous movement. Say to your- self at normal speed "Out and Pull". You should then have a drawn out centre section of around 6 inches in length. With tweezers, snap off 1 inch (2 cm) lengths of the thinned section (safety glasses!) and seal one end of each length with the Bunsen flame. Store in a sample bottle until needed.

Procedure Part 2

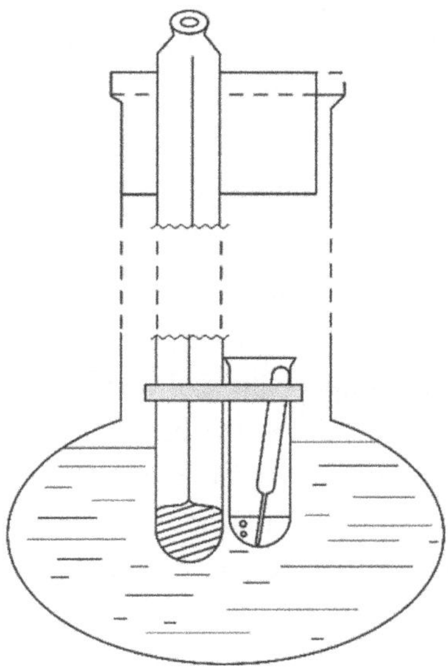

Obtain a *melting point* apparatus and a seal capillary tube at one end as you would for *melting point* determination. With tweezers, carefully place one of your specially drawn out tube *inside* the capillary tube so that the open end of the inner tube points towards the sealed end of the larger outer tube. Using a syringe, inject around 15 microlitres of the liquid whose boiling point you wish to determine into the capillary tube, ensuring that the syringe needle reaches down to the sealed end of the capillary. The assembly should appear as in the diagram. Do not add too much of the liquid.

Place the capillary tube in the melting point apparatus and heat it up as if you were finding a melting point. As the boiling point of the liquid is reached, a *continuous* stream of bubbles will appear from the open end of the inner tube. Immediately note the temperature (T_1) and stop heating. As the apparatus cools, note the temperature (T_2) at which the bubbles just *cease* to emerge from the tip of the inner tube. Both temperatures should be similar and denote the boiling point (if the difference is significant, the measurement obtained on cooling (T_2) is usually the most accurate).

N.B. One of the older style melting point apparatus can be used.

1.7 Vacuum Distillation (Distillation at Reduced Pressure)

High boiling liquids, or those which decompose at normal boiling point, are generally distilled at lower temperature under reduced pressure.

Use a round bottom flask as the distillation pot. The size of the flask depends on the volume to be distilled, but ideally it should be between one quarter and one half full – and no more. In order to prevent the *extremely bad bumping* that occurs with *every* vacuum distillation, the liquid should be stirred with a magnetic stirrer bar. The distillation pot is then fitted with a still head, quickfit thermometer (see demonstrator), condenser and a pig (2-way receiver adapter) with 2 round bottom flasks (see diagram below).

Place the liquid in the distillation pot and assemble the apparatus, all the ground glass joints should be lightly greased, with vacuum grease <u>not</u> Vaseline. Remember to support the collection flasks, until the system is under vacuum, gravity rules! Start stirring vigorously but do not heat. Turn the tap for the water pump on full before connecting the tubing to the system. Once connected, use the stopcock on the trap to carefully apply the vacuum. Ensure that the contents of the pot do not froth over and contaminate the collection flasks. If this happens, dismantle the apparatus and start again. Only when the apparatus is fully evacuated can the heater be switched on.

Therefore, if you purify a sample by distillation at reduced pressure, it is acceptable to quote the boiling range at which you collected the sample as long as you quote the pressure. With a water pump a pressure of 10-20 mmHg can be obtained, under these conditions, boiling points are reduced to approximately 100 ^{O}C.

Vigorous stirring is maintained throughout the distillation and fractions are collected by rotation of the pig at the appropriate point. When distillation is complete, turn of the heating and allow the apparatus to cool. Then, holding the collecting flasks, release the vacuum by opening the stopcock on the trap. <u>Do not swish the water pump off before releasing the vacuum.</u>

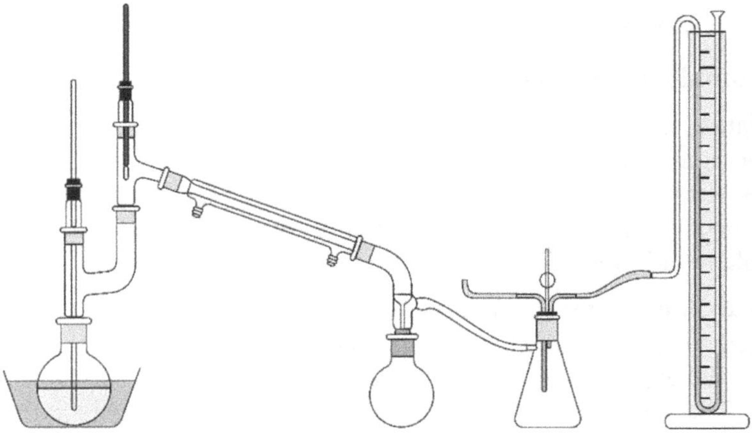

Vacuum distillation apparatus

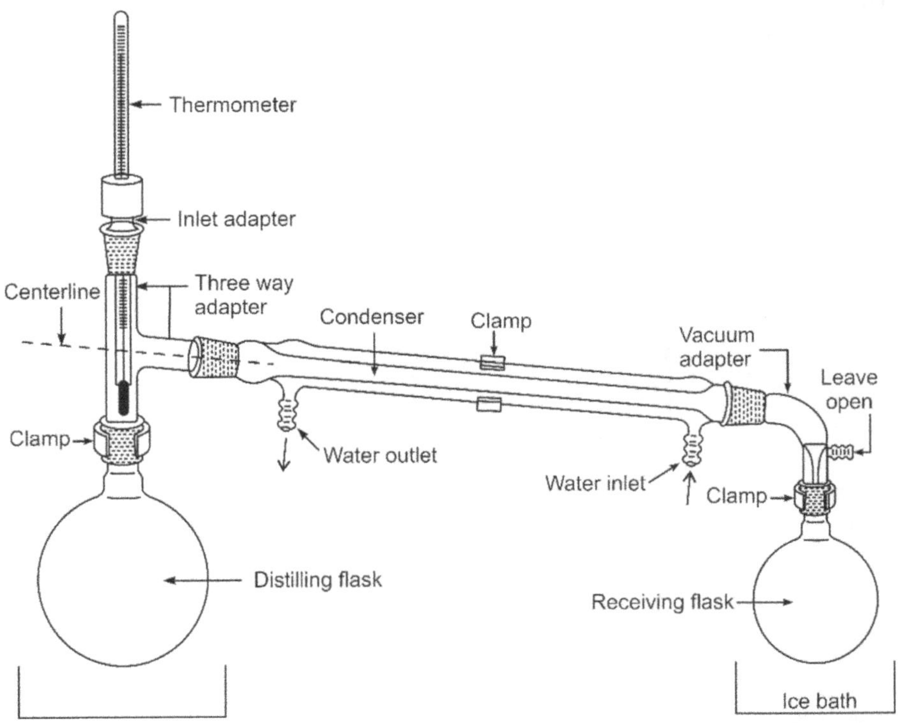

Simple distillation apparatus

1.8 Purification of a Liquid by Fractional Distillation

Fractional Distillation

Distillation and recrystallisation are the two chief methods of purifying organic compounds. Several kinds of distillations are possible, including simple distillation, fractional distillation, and steam distillation. This experiment is concerned with the separation, by fractional distillation, of two miscible liquids, methylene chloride and carbon tetrachloride.

Place 50 ml of the given mixture in the distilling flask and add a small boiling stone (to prevent bumping and to facilitate even boiling). Connect the flask to the apparatus and heat it gently with a small non-luminous Bunsen flame. The slower the distillation, the better will be the separation. Collect the 5 ml fraction of the distillate in a small measuring cylinder and note the temperature at which each fraction distils. Transfer each fraction to a clean, dry, numbered test-tubes; stopper the test-tubes and set them aside for subsequent g. l. c. examination.

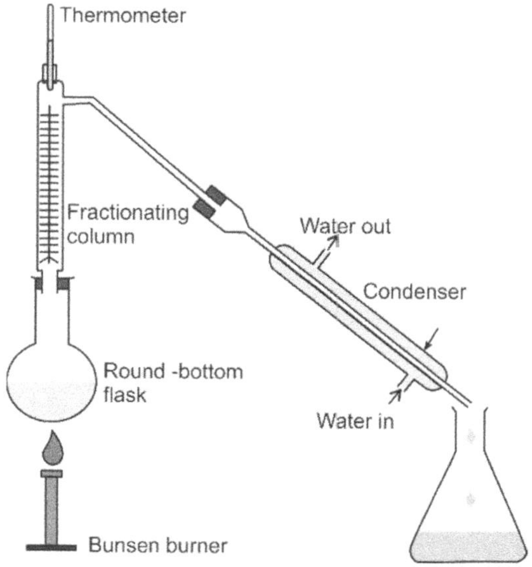

Fractional Distillation

1.9 Steam Distillation

The diagram for steam distillation is given below:

(i) It is a chemical process where a mixture which is made up of two or more components with different boiling points is separated.

(ii) The **mixture is heated until one of** the components boils (turns to vapor).

Steam is introduced into the distillation apparatus, **lowering the boiling points of the compounds**. The goal is to heat and separate the components at temperatures below their decomposition point.

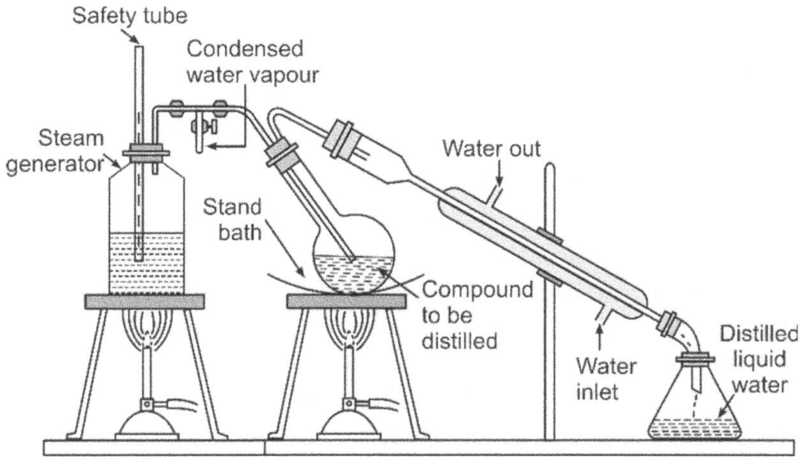

Steam Distillation

1.10 Recrystallisation

Purification of Organic Compounds: Solids by Recrystallisation

Crystallisation is an efficient method for the purification of solid organic substances. It is first necessary to find a suitable organic solvent, i.e. one which will readily dissolve the material when hot but only to a small extent when cold.

You will be required to recrystallise many of the compounds you prepared in order to purify them. Follow the simple procedure to ensure a good recrystallisation.

(a) Dissolve the given sample in the **minimum** amount of boiling solvent. Some insoluble impurities may be present. Ensure you do not add too much solvent trying to dissolve this portion. It is better to lose a little sample at this stage in order to get a very clean product.

(b) Filter the solution while hot to remove any insoluble impurities. This is usually done through a fluted filter paper. Filter the portions meanwhile keeping the remaining solution hot. If crystals form in the funnel wash them through with more hot solvent.

(c) Cool the filtrate and if necessary place in an ice-salt (NaCl) bath. The salt bath is prepared by mixing ice, a little water and some salt together. This will give a bath below 0 ^{o}C.

(d) Filter the precipitate by suction using a Karl Georg Büchner funnel to collect the purified product. The product may be washed with a little **cold** solvent. Continue to suck air over the crystals for a few minutes to help in drying.

1.11 Solvent Extraction

Separation Techniques in the purification of Organic Compounds

Solvent extraction is an operation very frequently used in practical organic chemistry. In this experiment one can apply this technique to the purification of a crude organic acid.

Many organic acids are sparingly soluble in water but are readily soluble in organic solvents such as benzene or methylene chloride. In contrast the sodium salts of these acids, which are obtained by the reaction of acids with aqueous sodium hydroxide, are readily soluble in water and insoluble in organic solvents. This is because the salts are ionic while the free acids are relatively un-dissociated.

This experiment makes use of these differences in solubility. You will be supplied, by your demonstrator, with a solution of an impure organic acid dissolved in benzene or any other organic solvent.

(i) **Formation of the sodium salt:** _Place the given solution (20 ml) in a separating funnel and add 20 ml of dilute sodium hydroxide solution. Close the funnel with a stopper and shake the contents thoroughly. Clamp the funnel vertically and allow the two layers to separate. The upper layer is benzene, the lower layer is sodium hydroxide solution. This treatment will have converted the organic acid into its water-soluble sodium salt which will be present in the lower alkaline solution. Non-acidic impurities will remain in the upper benzene layer.

Remove the stopper from the funnel and run the lower layer into a beaker. To make sure that all the organic acid has been removed from the benzene layer, re-extract this layer with a further 20 ml of dilute sodium hydroxide solution. Combine the two alkaline solutions in the beaker and discard the benzene solution in the funnel.

(ii) **Regeneration of the Free Organic Acid:** Add dilute hydrochloric acid with stirring to the alkaline solution in the beaker until the mixture is just acid (test with Congo Red indicator paper which turns blue in acid solution). This treatment will cause most of the organic acid to be precipitated as a white solid. Pour the resulting suspension of the organic acid into a clean separating funnel.

(iii) **Re-extraction of the Organic Acid:** Add methylene chloride (15 ml.) to the suspension in the separating funnel, stopper the funnel and shake the contents thoroughly. Allow the layer to separate. The organic acid will now have dissolved in the lower methylene chloride layer. N.B. Methylene chloride, unlike benzene is denser than water and hence forms the lower layer at this stage. Run off the lower layer into a beaker. Re-extract the upper layer remaining in the funnel with a further 15 ml. of methylene chloride. This should extract the last traces of the organic acid from the aqueous layer. Combine the two methylene chloride solutions in the beaker and discard the aqueous layer in the funnel.

(iv) **Purification of the Extract:** Pour the methylene chloride solution of the organic acid into a clean separating funnel, add 20 ml. of cold water and shake the mixture thoroughly. Traces of inorganic salts will dissolve in the water. Remove the stopper, allow the layers to separate and run the lower methylene chloride solution into a small, clean dry conical flask. Add a few grams of solid anhydrous sodium sulphate to the methylene chloride extract and after swirling the flask, allow it stand for 10 minutes. This removes the traces of water from the organic solution.

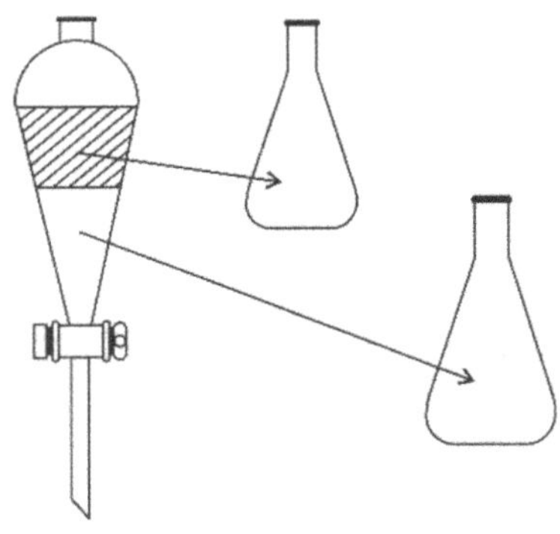

(v) **Isolate of the Pure Organic Acid:** Decant the clear methylene chloride solution into a small round-bottomed flask. Rinse the remaining sodium sulphate with a small additional amount of methylene chloride and decant the washings into a round-bottomed flask. Add a small piece of carborundum (boiling stone) and remove the methylene chloride solvent by distillation on the steam bath using the simple distillation apparatus. A residue of the purified organic acid will remain in the round-bottomed flask.

Recrystallise the material from the minimum amount of boiling water. Collect, dry and weigh the crystals. Determine the melting point of the pure dry organic acid and place the crystals in a labelled sample tube.

CHAPTER 2

Chromatography

2.1 Thin Layer Chromatography (TLC)

The use of this technique can be demonstrated by analyzing a dye mixture and a two-component mixture.

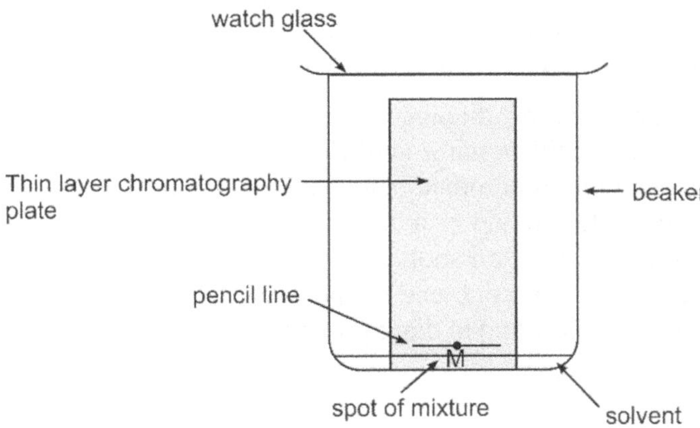

Thin layer chromatography

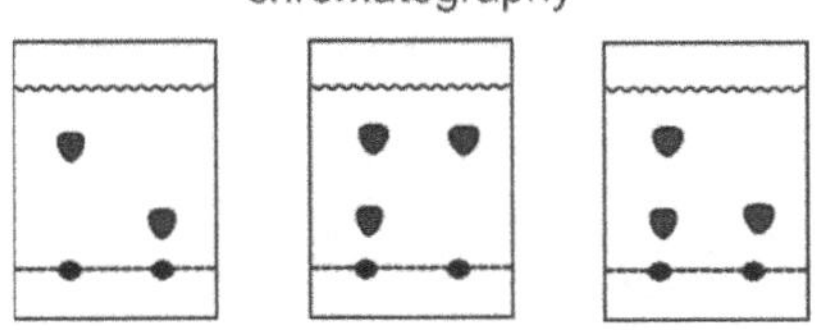

Object: To separate and identify the components of a mixture by reference to known substances

Background: Chromatography, of which there are many forms, is a technique for the separation of a mixture and purification of compounds. Thin layer chromatography (t.l.c) is a simple and convenient form of this technique enabling both micro and preparative quantities of mixtures to be separated rapidly.

Essentially a glass, metal or plastic plates are covered with a uniform thin film absorbent, e.g. alumina or silica gel-G. The film is activated by drying, and a solution of the mixture to be separated is spotted on to the plate, which is then placed upright in a bottle containing a solvent. The solvent(s) ascends the absorbent film by capillary action, carrying with it the components of the mixture at different rates depending on their affinities for the adsorbent. The separated components, if not colored, are then made visible by placing the chromatography plate in a bottle or jar containing iodine. Alternatively the chromatogram can be examined in ultraviolet light.

2.1.1 Experimental Procedure

Dye mixture – Initially, practice the use of TLC using the mixture of dyes (provided red, yellow, blue). Use toluene as the developing solvent.

Application of sample - Make an applicator by softening the middle of an unsealed melting point tube over a very small flame (from a micro burner) and quickly draw the ends of the tube apart. When it is cool, snap the capillary in half to make two applicators.

Fill the applicator by dipping the capillary tip vertically into the dye mixture solution. Still holding applicator vertical, touch the capillary tip onto a horizontal micro-chromatoplate (supplied). Gently blow over the spot during the application to aid evaporation of the solvent and to keep the spot within 3 mm diameter. Two spots per plate may be applied. Apply them so that they are 5 mm from one end of the plate, 8 mm apart and equidistant from the edges, as indicated in diagram 1 below.

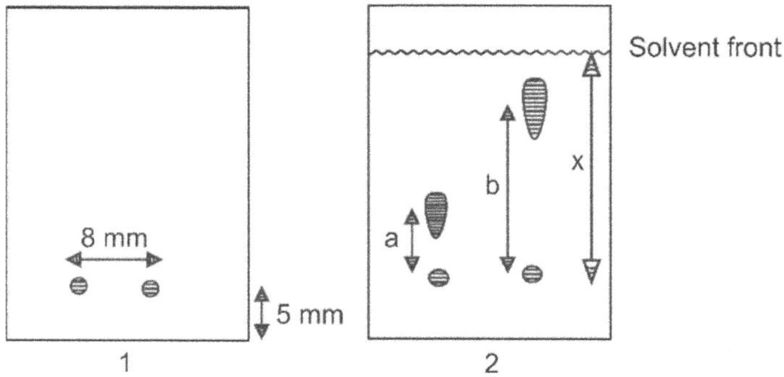

Separation of dye mixture

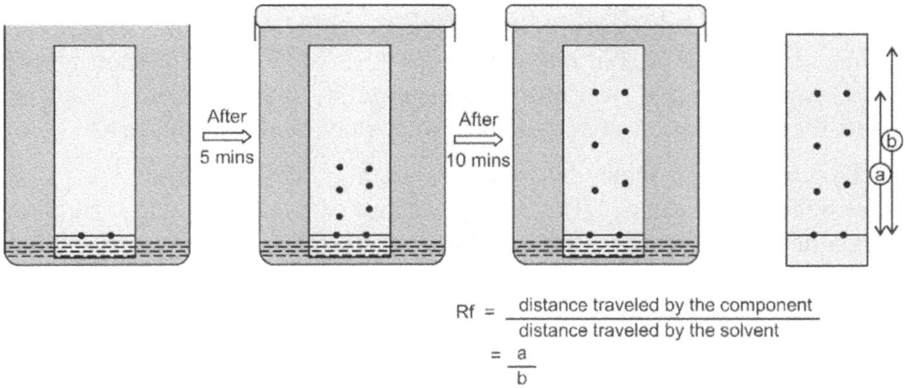

Steps of thin layer chromatography

2.1.2 Development of Plates

Line the inside of a 250 ml beaker with a strip of filter paper, then pour in the developing solvent to moisten the paper and give a depth of 2-3 mm in the bottom of the beaker. Place chromatoplate upright in the beaker (with the end with the spots on at the bottom). Cover the beaker with a watch glass. When the solvent has risen almost to the top of the plate (5-6 cm.), remove the plate from the beaker and immediately mark the position of the solvent with a pencil. Allow the plate to dry. You should see sharp and well separated colored spots corresponding to the components of the dye mixture. Initially over loading the plate with too much dye mixture will give poor resolution of the spots.

2.1.3 Visualization of Plates

There are many ways to visualize the spots on an analytical t.l.c. plate. They can be roughly split into two types: those which interact reversibly with the compounds to be detected and those which interact destructively. The two most common methods are examples of the former: U.V. and iodine.

The silica on the plates used contain a fluorescent additive (GF_{254}). When viewed under U.V. light with wavelength of 254 nm (short wave lamp) the plate has a green fluorescence. A compound which absorbs the U.V. radiation, will prevent the fluorescent material receiving electronic excitation, and thus appears as a dark spot on the green background. The easiest way to mark the positions of the spot is to draw around them with a pencil. Alternative visualization procedures are available for use in analytical t.l.c., although U.V. visualization is probably sufficient for most of the t.l.c.'s you will run. The U.V. lamps are housed in a closed box. Under no circumstances should you look directly at the lamp or expose unprotected skin to the U.V. radiation.

Most compounds form colored complexes with iodine. There are iodine/sand jars available on each bench (provided by the demonstrator), these are easy to use- just push the plate gently into the iodine /sand jar, leave for a minute or so and pull out. Note which spots (if any) react.

Examples of methods which interact destructively are: 2,4-dinitrophenylhydrazine (Brady's reagent) for visualizing carbonyl compounds. Vanillin for alcohols and acids, aqueous potassium permanganate for compounds which can be oxidized (alkenes, alcohol etc) and ninhydrin for amino acids. These can be applied as sprays or as 'dips" (the whole plate is immersed in the reagent for a few seconds before being dried).

Once the spots have been visualized the R_f values should be calculated and presented in the report.

R_f = Distance travelled by spot from origin / Distance travelled by solvent from origin

Please note this is not an absolute value, since each plate is unique there will be slight variations in the R_f value.

2.1.4 Choice of Solvent

A high R_f is obtained if the compounds have a greater affinity for the solvent than for the adsorbent (normal silica has hydrogen groups on its surface, i.e. it's a polar). The more polar the solvent the higher the R_f for a given compound. Commonly used solvents are listed below in order of decreasing polarity:

Acetic acid < Methanol < Ethanol < Acetone < Ethyl acetate < Dichloromethane < Diethyl ether (ether) < Toluene < Hexane (Petroleum ether)

In general, mixed solvent systems are used. These are chosen so as to give R_f values between 0.3 and 0.7 for optimal resolution.

2.1.5 Analysis of an Unknown Mixture

When a satisfactory separation of the dye mixture has been achieved, obtain a solution of two-component mixture from the demonstrator. This will contain one pair from the materials A, B, C and D in solution in toluene. Solutions of the individual compound required as reference are available in the laboratory, supplied by the demonstrator. For each of A, B, C and D place spot of the reference sample and a spot of the unknown mixture on four separate chromatoplates. A different applicator must be used with each reference sample. Dry the plates in the oven. Develop each plate with 5%

ethyl acetate/toluene already prepared by the demonstrator, and allow to dry, first in the air and then again in the oven. Visualize the compound by the UV Chromatoview, use a pencil to draw an outline of the visualized spots. In the absence of UV chamber, all spots can be visualized by placing the dried plates in an iodine/sand jar.

2.1.6 R_f Values

Mark the position of the centre of the spots using a pin and for each spot calculate the approximate R_f of the compound.

Calculate the R_f values of each of the pure dyes and for the components of the mixture using the relationship

$$R_f = \frac{\text{Distance travelled bythe spot}}{\text{Total distance travelledby solvent}}$$

Thus the R_f of the two compounds in diagram 2 above are

$R_f = a / x$ and $R_f = b / x$

It should then be possible to identify the unknown components of your mixture by comparing R_f with those for the reference materials.

2.1.7 Confirmation

Finally confirm your conclusions by running a further chromatogram using the unknown mixture as one spot and the suspected mixture as a second spot on the same plate. This procedure eliminates the variation in R_f value caused by non-reproducible performance of different chromatoplates.

In general, in choosing a developing solvent so that a separation of the components of a mixture is achieved, it is usual to start with a relatively non-polar solvent like petroleum ether. If the compounds do not move on the chromatoplate use a more polar solvents mentioned above. Frequently a mixed solvent is valuable. Thus if the compound moves only slightly with toluene, try a 9:1 mixture of toluene & dichloromethane. Conversely, if you find that the compounds move too far up the plate, so that a poor separation is achieved, use a less polar solvent.

2.1.8 Results

Fill in the tables below with your final measurements

Distance travelled (mm)

Plate No.	1	2	3	4
Solvent				
Standard	A	B	C	D
Unknown X				
Unknown Y				

R_f values

Standard	A	B	C	D
Unknown X				
Unknown Y				

Conclusion: X = Y =

2.2 Paper Chromatography

The technique was such a remarkable advance upon previous methods, and of such wide application, that its discoverers, Martin and Syringe, were honored by the award of Nobel prize for chemistry in 1952.

This is usually regarded as a typical partition system, where the stationary phase is water, held by adsorption on cellulose molecules, which in turn are kept in a fixed position by the fibrous structure of the paper. It is now realized, however, that adsorption of components of the mobile phase and of solutes, and ion exchange effects, also play a part and that the role of the paper is by no means merely that of an inert support.

One characteristic feature of the paper technique is that there is nothing corresponding to the column effluent of the usual gas or liquid system. The separated substances are located and identified on the paper and there is thus a reasonably permanent record of the separation. There is no collecting of a series of fractions and sophisticated continuous monitoring devices are not needed. Quantitative determination of the separated substances may be carried out on the paper, but it is desired to remove them by cutting out the part of the paper containing each substance and elute each one separately.

Paper chromatography is a form of partition chromatography especially adapted to separating very small quantities of material. The cellulose of filter paper contains adsorbed water (about 20%). This combination is the stationary phase. A drop of a solution containing the mixture to be separated is applied in solution as a spot on the filter paper, and the developing solvent (which is immersible with water) passes through a filter paper by capillary flow. The component of the sample will move at different rates through the paper. The rates being proportional to the partition coefficients between the adsorbed water and the particular solvent used for developing, although adsorption by cellulose is also important.

When the position of the separated zones have thus been detected, it is essential to identify each individual component. In the normal course each one gives a characteristic color with the locating agent, as is often found with inorganic substances, but less commonly with organic compounds.

The simplest method of identification is based on the R_f value, the ratio of the distance travelled by the spot and that travelled by the solvent front.

Paper chromatography is an invaluable tool in biochemistry, especially for separating and identifying mixtures of amino acids.

Experiment

Separation of synthetic food dyes: Using the apparatus shown below, a spot of a dye solution consisting of the mixed dyes in methanol is applied near the centre of a piece of filter paper punctured at the centre. A filter paper wick is placed through the central hole; the filter paper is placed over the solvent contained in the lower half of a petridish, which is covered to reduce evaporation. As the solvent travels up the wick and through the paper, the separate components appear as concentric rings

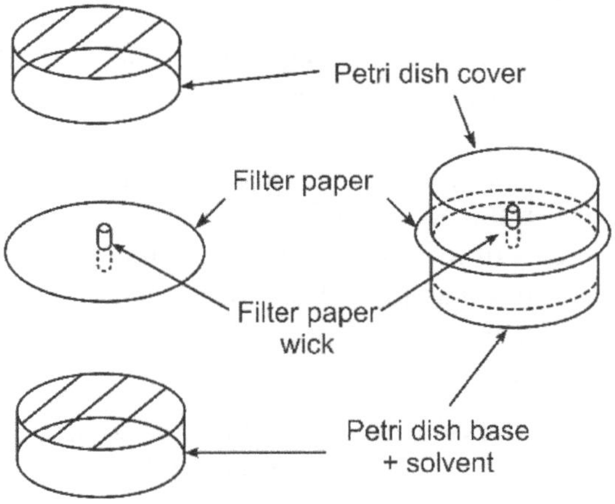

2.3 Column Chromatography

This procedure utilizes differences in the strength of adsorption of given compounds to the surfaces of solids. These attractions are very dependent upon the nature of the solid adsorbent as well as the substances adsorbed. Since common adsorbents, such as alumina, silica gel, silicic acid and cellulose, are reasonably polar in character, the more polar compounds are more strongly adsorbed.

In this type of chromatography the separation is effected by adsorption on a solid stationary phase. Several types of forces that cause interaction of the substance with the solids are hydrogen bonding, vanderwaals forces, electrostatic interaction etc. A cylindrical glass column of the type shown below (page – 30) is employed to separate a mixture by column chromatography. The size of the column which should be used depends on the volume of the sample to be separated. For many purposes a bed volume of about five times the sample volume should be suffice, with a ratio of diameter to height of about 1:15, although for very intricate separations the ratio may need to be as much as 1:100.

To pack a column, first clean the glass column thoroughly and dry. Now position it vertically on an iron stand with the help of a big clamp. Close the stop cock and fill the column with an appropriate solvent. Insert a small plug of glass wool to the bottom of the column. Pour some fine dry sand so that it forms a 1 cm layer on the top of the glass wool. Then the column is packed with required quantity of a suitable adsorbent; while the liquid is slowly draining from the pipette the tip is run round the inside of the column, care being taken not to touch the packing; any sample remaining on the walls can be carefully washed on to the column in a similar way, using pure solvent. Then a suitable solvent (mobile phase) is allowed to percolate through the column. The process is known as elution. It is common to use a relatively non-polar solvent in the beginning. The adsorption depends on both the nature of the solvent and the adsorbent. For effective separation the eluting solvent must be significantly less polar than the components of the mixture. The rate of the movement of a given component is related to the strength of its adsorption to the solid. Different compounds vary in this respect and if the rates of movement on the column are somewhat different, a separation will occur. Polar solvents are used for eluting strongly adsorbed components while non-polar solvents are used for weakly adsorbed components of a mixture.

An approximate order of increasing polarity of some common solvents is as follows: Petroleum ether, hexane, benzene, toluene, chloroform, ethanol, water.

A mixture of two solvents of different proportions may be more useful for separation than a single solvent in many instances.

The most widely used adsorbents in adsorption chromatography are aluminium oxide, (alumina) and silica gel. The particle size of the commercially available grade is in the range 50-200 μm (70-290 mesh). Alumina may be obtained in basic (pH 10), neutral (pH 7) and acidic (pH 4) forms and it is important to ensure the correct type is employed. For example, basic alumina may lead to hydrolysis of esters, acidic alumina may lead to dehydration of tertiary alcohols; in these circumstances neutral alumina is to be recommended.

Experiment 1

2.3.1 Separation of a Mixture of o- and p-nitroaniline

Aim: To separate o-nitro aniline and p-nitro aniline in the given mixture

Materials needed: Chromatography column (3 cm internal diameter, 30 cm length), silica gel, (column grade, 60-120 mesh, 200 g.), conical flasks, mixture of o-nitro-aniline, and p-nitro aniline (100 mg each), cotton, t.l.c. micro slides (for monitoring the column chromatography separation).

Procedure: Clean the column with chromic acid. Rinse with distilled water and dry it. Loosely plug the bottom end of the column with a small piece of cotton using a glass rod. Fill the column with hexane/petroleum ether to half level. Prepare uniform slurry of 75 grams of silica gel in about 300 ml of benzene. Carefully transfer the slurry into the column with a stopcock slightly open. The slurry slowly settles down. Drain out any excess solvent. When all the solvent is drained out and the slurry level in the column is unchanged, close the stopcock. The packed column height should be 2/3 of the length of the column. Now, clamp the column vertically and carefully. Load the mixture of o-nitro aniline and p-nitro aniline as discrete plug by adsorbing it over silica gel. Then insert a further small loose plug of cotton over the sample. Moisten the column by running in 25 ml hexane from the dropping funnel; the column must be kept covered with solvent to a depth of about 5 cm. throughout the experiment to avoid the formation of air bubbles. Open the stopper and fill the column with 10% ethyl acetate/hexane. Feed the solvent continuously from the top. After complete elution, initially an orange band appears at the top of the column.

After 30 minutes two bands, a clear orange band (due to o-nitro aniline) moving faster than the pale yellow band (due to p-nitro aniline) with a wide gap in between them appear. Collect the column eluate in serially labeled conical flasks. In each conical flasks collect about 150 ml eluate. Concentrate these fractions separately to a small volume. Examine their TLC. Combine the fractions that show the same t.l.c. behavior and evaporate the solvent to get pure o-nitro aniline and p-nitro aniline. Record the yield and their melting points.

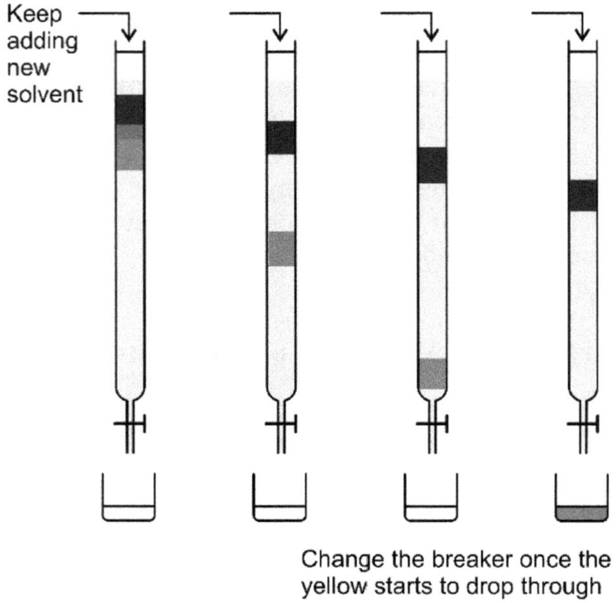
Keep adding new solvent
Change the breaker once the yellow starts to drop through

CHAPTER 3

Qualitative Organic Analysis

3.1 The Possible Combination of Binary and Ternary Mixtures

***Examples of binary mixtures*:**

1. Strong acid + Neutral including Hydrocarbons
2. Strong acid + Weak acid
3. Weak acid + Neutral including Hydrocarbons
4. Base + Hydrocarbon
5. Carbohydrate + Weak acid
6. Carbohydrate + Strong acid (other than p-nitro benzoic acid)

Organic mixture analysis:

Examples of ternary mixtures:

1. Strong acid + Weak acid + Neutral
2. Carbohydrate + Weak acid + Neutral
3. Base + Hydrocarbon + Weak acid
4. Carbohydrate + Strong acid (other than p-nitro benzoic acid) + Neutral

3.2 Identification of Unknown Organic Compounds

You will be provided with a sample of an unknown compound together with its infrared, ^{1}H NMR ^{13}C NMR spectra (if department provides the spectra) and you will be informed which elements other than C, H, or O it possesses, if any. You should use a combination of chemical tests to identify the unknown compound fully, finally confirming your conclusions by preparing a suitable crystalline derivative.

The correct procedure begins with the identification of the functional groups present in the compound. Table I lists those which are likely to encounter in the organic laboratory.

Table I Commonly encountered functional groups.

Class of Compound	Functional Group
Alkene	>C=C<
Arene	Benzene ring
Alcohol (primary, Secondary or tertiary)	-OH bound to a saturated C atom
Phenol	-OH bound to the C atom of a benzene ring
Ether	$R-O-R^1$; two alkyl or aryl groups bound to an O atom
Aldehyde	RCHO; carbonyl group bound to an H atom and an alkyl or aryl group
Ketone	$RCOR^1$; carbonyl group bound to two alkyl or aryl groups
Carboxylic	-COOH (or-CO_2H)
Ester	-COOR; H atom in a carboxylic acid replace by an alkyl or aryl group
Amide	-$CONH_2$, -CONHR, or – $CONR_2$
Acid anhydride	-CO-O-CO-
Nitrile	-CN
Amine (primary aliphatic) Amine (primary aromatic)	-NH_2 bound to a saturated C atom -NH_2 Bound to a benzene ring
Nitro compound	-NO_2 bound to alkyl or aryl group
Halide	-F, -Cl, -Br or –I bound to a C atom

Procedure: Work through the tables and carry out the tests in the correct order

3.3 Preliminary Tests

Examine the infrared spectrum of your unknown compounds and make what deduction you can as regards the nature of any functional groups it possesses. Take note of what you know about the presence of elements other than C, H, and O.

Carry out the preliminary tests (table II) to obtain further information as to which functional groups are present. Use small quantities of your unknown in test tubes for these tests.

Consult table III and carry out any relevant spectroscopic search or supplementary test.

Check your conclusions against the information in Table IV to VI and use your spectroscopic textbook or other relevant information from your spectroscopy course.

N.B. Nitrogen containing compounds will be amines (soluble in dilute acid), amides (infrared band at around 1650 cm^{-1}), nitriles (infrared band at around 2250 cm^{-1}), or nitro compounds (usually reduce solutions of titanium (III) chloride, infrared band at 1550 and 1350 cm^{-1}).

3.4 Identification

Having determined the functional groups present, consult the tables of physical constants (Table VII to XVI) and tentatively identify the compound.

Examine carefully the ^{1}H NMR and ^{13}CNMR spectra to see if your tentative assignment of structure is consistent with the spectra. Change your assignment if necessary.

Make a crystalline, solid derivative of your unknown sample. Choose a derivative which has a melting point significantly different from that of derivatives of other possible compounds.

Make a definite identification of your compound.

3.5 Report

Present a complete report and submit a sample of the derivative(s) in a correctly labeled sample tube. A specimen report, for your guidance, is given at the end of this chapter.

Note down your observations and do as much writing up as possible while you are in the laboratory.

Aim at a reasonable standard of neatness, but remember that accuracy is promoted by making immediate notes of your observations.

Always write in ink – not pencil – cross out errors neatly, rather than covering them up with Tipp-Ex or similar fluids.

3.6 Tables of Preliminary Tests for Functional Groups and Spectroscopic Data

The following table contains the preliminary tests for functional groups **(Table II and III)**, which you should carry out in the order given. Pay attention to the notes which accompany Table I.

Check your deductions against the data in Table IV to V. However, note that the spectroscopic information in these tables is not comprehensive – just a summary of the most important features.

Once you have identified at least one functional group in your unknown sample proceed to make a tentative identification using the tables of physical data (melting and boiling points) in section **3.7** of this manual.

Table II Preliminary tests for functional groups.

Test	Observation	Inference
Note appearance and smell	Colored	Probably unsaturated (see note i)
Determine m.p. or b.p.		*(note ii)*
Test solubility in water (cold then hot).	Compound is water soluble	Polar compound, e.g. alcohol, carboxylic acid
Test pH of the solution or suspension in H_2O	pH < 7	Carboxylic acid *(note iii)*
Test solubility in dilute. HCl	More soluble in water	Amine
Test solubility in dil. NaOH	More soluble than in water	Acidic compound – phenol or carboxylic acid (*note iv*)
If the compound is soluble in dil. NaOH, test its solubility in $NaHCO_3$ solution	Soluble in NaOH and $NaHCO_3$ Soluble in NaOH but not $NaHCO_3$	Carboxylic acid (*note v*) Phenol
If compound contains N but is not basic, heat with dil. NaOH.	Dissolves with evolution of NH_3	Primary amide
Treat compound separately, drop-wise with Br_2/CCl_4 and Br_2/H_2O	Rapid decolorisation with either reagent	Unsaturated compound – alkene, or one that is easily brominated. *(note vi)*
Treat compound with aq.$KMnO_4$. If no change warm on a steam bath.	Decolorisation cold Decolorisation on warming	Unsaturated compound alkene easily oxidizable group
Add 1-2 drops of liquid (or 20mg of solid dissolved in ca. 0.5ml methanol) to Brady's reagent (1ml).	Formation of orange or red precipitate	Aldehyde or ketone (*note vii*)
If Brady's test positive, test with Tollen's reagent. (If no reaction after two min., warm mixture on steam bath).	Silver mirror or black precipitate formed	Aldehyde (*note viii*)
If compound contains N, boil it with $TiCl_3$ solution.	Decolorisation	Nitro compound *(note ix)*

3.6.1 Notes for Table II

Note (i): In some cases the smell of unknown compound can be helpful guide to its identity the following are examples:

Like phenol : simple phenols

Like aniline : simple aromatic amines

Fruity odour : simple esters

Pungent : Aldehydes, lower aliphatic carboxylic acids and anhydrides

Great caution should be exercised to smell unknown compounds. In these cases smelling is hazardous and therefore bad practice.

Note (ii): The compounds supplied as unknown are generally commercial samples of at least reasonable purity. As such they should melt or boil over a narrow range, but may have melting or boiling points which differ by a few degrees from those given in tables.

If a solid melts followed by decomposition, e.g. becomes darker or gives off vapour, record as e.g. m.p. 235 – 238 ^{O}C: If it decomposes before melting, record as e.g. "dec. 235 ^{O}C". A method for determining boiling points is described on the next page.

Note (iii): Interpret the pH test with caution. An acidic impurity, especially a carboxylic acid in an aldehyde, can give a misleading result, but this will be revealed in later tests.

Note (iv): When a carboxylic acid dissolves in aq.$NaHCO_3$, bubbles of CO_2 are frequently observed. Look for this and record it if observed.

Note (v): If a phenol is suspected, attempt to confirm this by treating the compound with $FeCl_3$ solution. A distinct color change indicates a phenol, but the test is not always positive, even when a phenol is genuinely present.

Note (vi): Bromine is decolourised by addition to alkene carbon-carbon double bonds, and also by bromination of certain other types of compounds, particularly phenols and amines. Alkenes with electron withdrawing groups, e.g. – CO_2R, will react relatively slowly. Interpret your observations in the light of other tests.

Note (vii): Amines sometimes give an apparently positive test with Bready's reagent, by simply reducing the acidity and precipitating the weekly basic reagent, 2,4-dinitrophenylhydrazine.

Note (viii): Tollen's reagent is relatively unstable and must therefore be freshly prepared each time it is needed. Add 1 drop of dilute NaOH solution to 2ml of 2% aq. $AgNO_3$. Where by grey coloured AgO precipitate forms. Then add dil. ammonia drop wise until the precipitate just dissolves. Add 1 drop or a few crystals of the unknown to this solution. If silver mirror or black precipitate does not develop within a minute or two, warm the mixture on a steam bath. Dispose of the reagent as soon as the test has been completed. Dissolve any silver mirror in dil. nitric acid and dispose of the solution down the sink with much water.

This is difficult test, which frequently gives negative or inconclusive results, even when an aldehyde is genuinely present. Adding too much ammonia during preparation of the reagent can cause problems, as can the presence of acid impurities in the aldehyde. Another mistake is to warm the test mixture too strongly, allowing a volatile aldehyde to boil off before reacting.

Note (ix): Not all nitro compounds decolourize Ti (iii) solutions. Look for other evidence of a nitro group, e.g. colour and characteristic infrared absorptions.

Table III Supplementary tests for functional groups.

<table>
<tr><td colspan="3">The tests in table II should reveal the presence of Alkene, Phenol, Aldehyde, Ketone, carboxylic acid, Amine and possibly also primary amide, nitro group.</td></tr>
<tr><td colspan="3">If you have already identified a functional group in your unknown sample, check your infrared, and NMR spectra for the expected characteristic features listed in table IV-VI. If there is no primary amine group present, proceed to a tentative assignment based on Tables VII-XVI.</td></tr>
<tr><td rowspan="2">If a primary amine group present Dissolve or suspend 0.1g of the compound in aq. HCl (2ml of conc. acid + 3ml H_2O), and cool in ice to < than 5^0. Slowly add 10% aq $NaNO_2$ until excess nitrite is present (test with starch – KI paper). It is important to keep the temp. < 10^0 during the addition.
If there is a little or no evolution of N_2, test with starch – KI paper for excess nitrite and remove any by addition of urea (N_2 evolved). Add 2ml of the solution to a solution of 2-naphthol (50mg) in dil. NaOH (2ml).
Now proceed to make an assignment based on tables VII-XVI.</td><td>Observation</td><td>Inference</td></tr>
<tr><td>Evolution of N_2, continuously and vigorously, while solution remains clear.
Intense orange or red dye produced</td><td>Primary aliphatic or side-chain aromatic amine.
Primary aromatic amine</td></tr>
<tr><td colspan="3">If you have not yet found a functional group, consider other groups like alcohol, ester, amide, nitrile.
These four functionalities have characteristic infrared absorption and, in some cases NMR signals, which are listed in Tables IV and V. Check the infrared and NMR spectra of your unknown carefully for the expected features of each of these possible functional groups.</td></tr>
<tr><td rowspan="2">If an alcohol is suspected
Dissolve 2-3 drops of the unknown in water (1ml). Add this to 1ml of cerium (IV) reagent.</td><td>Observation</td><td>Inference</td></tr>
<tr><td>Red complex formed</td><td>Alcohol</td></tr>
<tr><td colspan="3">If you have still not identified a functional group, you will have to repeat some or all of the tests in Table II, and give more consideration to your spectra.</td></tr>
</table>

Table IV Functional group characteristics. Groups not containing nitrogen.

Functional group	*IR* (ν_{max}cm^{-1})	^{1}H NMR (δ_H/ppm)	^{13}C NMR # (δ_C/ppm)	Characteristic chemical reactions
Alcohol	-3300 v(0-H)	2.5-5.0 (broad, variable)	--	Orange/red color with Ce(IV) ammonium nitrate
Phenol	-3300 v(O-H)	4.5-10 (broad variable)	--	Soluble in aq. NaOH, but not $NaHCO_3$
	2820 – 2720 v(C-H) RCHO 1725cm^{-1}(CO) ArCHO 1690 cm^{-1} CO	-C<u>H</u>O 9.7-10.0	-<u>C</u>HO 200-205(*d*)	1. Orange/red ppt. with Brady's reagent 2. Ag mirror with Tollen's reagent
Ketone	R_2C=O 1715 v(C=O) ArCOR 1690 v(C=O)	------	R_2<u>C</u>O 205-215 (*s*) Ar<u>C</u>OR 195-200 (*s*)	1.Orange/red ppt.with Brady's reagent, 2. No silver mirror with Tollen's
Carboxylic acid	3300-2700 v(O-H), very broad 1710 v(C=O) broad	-COO<u>H</u> 12-15 (broad)	-<u>C</u>OOH, 165-180 (*s*)	Soluble in aq. NaOH and $NaHCO_3$
Ester	RCO_2R, 1735 v(C=O) 1240 v(C-O) $ArCO_2R$ 1720 v(C=O) 1280-1120 v(C-O)	>C<u>H</u>-O-COR 3.5-5.0	R<u>C</u>O_2R 170-175 (*s*) Ar<u>C</u>O_2R 165-170 (*s*)	---------
Acid anhydride	(RCO2)O $(RCO_2)O$ 1820 and 1760 Maleic anhydride and phthalic 1850 and 1780 v(C=O)	--------	- -<u>C</u>O-O-<u>C</u>O- 165-175 (*s*)	Soluble in aq. NaOH on warming Can be confused with carboxylic acid

Table V Functional group characteristics. Groups containing nitrogen.

Functional group	IR (ν_{max}/cm^{-1})	^1H NMR (δ_H/ppm)	^{13}C NMR # (δ_C/ppm)	Characteristic chemical reactions
Amide	2 bands -1650 ν(C=O)	-$CONH_2$ 5-8.5(often very broad	R (Ar)$CONH_2$ 165-180 (*s*)	Primary amides ($RCONH_2$) give off NH3 when heated with aq. NaOH.
Nitrile	-2250 ν(CN)	-----	R(Ar)CN 115-125 (*s*)	-------
Amine (primary)	2 bands -3300 ν(N-H)	-NH_2 3.5 – 8.0 (variable, broad)	----	Soluble in dil. HCl
Nitro compound	2 bands -1550-1350 ν(NO)	-----	------	Most decolourise $TiCl_3$ solutions on warming

Table VI Functional group characteristics.
Groups which will be present only when accompanied by others.

Functional group	IR (ν_{max}/cm-1)	^1H NMR (δ_H/ppm	^{13}C NMR # (δ_C/ppm)	Characteristic chemical reactions
Alkene	-CH=CH 1650-1640 ν(C=C) -CH=CH-CO- -1600 ν(C=C)	-C<u>H</u>=C- < 5.0-6.0	><u>C</u>=<u>C</u>< 165-175 (*s,d* or *t*)	Decolourises Br_2 reacts with $KMnO_4$
Arene	-1600, -1525 ν(C=C) C_6H_5X 750 and 700 o-C_6H_4XY 750 m-C_6H_4XY 830 and 770 p-C_6H_4XY 830	6.5-8.0 1,4-disubstitution usually results in a symmetrical pattern of signals.	110-160 (*s* or *d*)	-----------
Ether	------------	>C<u>H</u>-O-R 3.3 - 4.0	--------	-------
Halide	------------	----------	--------	You may expect unknown contains F, Cl, Br, I.

3.7 Table of Physical Constants of Compounds and their Derivatives

Once you have identified a functional group in your unknown sample, use the melting or boiling point data given in the tables in this section to make a tentative identification. Then chose a crystalline derivative for final confirmation.

Ideally a derivative will have a melting point above 100^0 C and below about 250 ^{O}C, and the melting should distinguish clearly amongst all the possible alternatives. Some derivatives are difficult or impossible to crystallize, can occur in more than one crystal form, with different melting points, or otherwise not recommended. Therefore, if a melting point is not given for a particular derivative, do not try to prepare it. If a proposed derivative is a liquid (b.p. rather than m.p. quoted), it is completely unsuitable.

Here is a list of the tables in this section of the manual.

3.7.1 Tables

VII. Alcohols, **VIII.** Aldehydes, IX. Phenols, X. Ketones, XI. Carboxylic acid, XII. Esters, XIII. Anhydrides, XIV. Amides, XV. Nitriles, XVI. Amines

If your unknown compound has more than one of these functional groups, it may be found in more than one of the tables.

The experimental procedures for making the derivatives of unknown samples are given in this section.

Table VII Alcohols.

Melting points of derivatives					
Alcohol	**M.p.**	**B.p.**	**3,4-Diniro-benzoate**	**4-Nitro-benzoate**	**Hydrogen 3-nitrophthalate**
Methanol	---	65°	107°	95°	---
Ethanol	---	78	94	56	161-162°
Propane-2-ol	---	83	122	108	151-7°
t-Butanol	25°	83	122	116	---
Propane-1-ol	---	97	73	34	139-40°
Butane-2-ol	---	100	76	----	---
2-Methylbutane-2-ol (t-amyl alcohol)	---	102	118	85	---
Isobutanol	---	108	87	69	178-82°

Table VII ***contd...***

Melting points of derivatives					
Alcohol	**M.p.**	**B.p.**	**3,4-Diniro-benzoate**	**4-Nitro-benzoate**	**Hydrogen 3-nitrophthalate**
Butane-1-ol	---	117	63	35	147
Cyclopentanol	---	140	120	62	---
Cyclohexanol	25	161	113	51	----
Benzyl alcohol	---	205	113	85	178
Diphenylcarbinol	69	---	141	131	----

N.B. Some melting points show wide ranges due to hydrate formation.

Table VIII Aldehydes.

Melting points of derivatives					
Aldehyde	**M.p.**	**B.p.**	**2,4-Dinitro-phenylhy drazone**	**4-Nitro-phenylhy drazone**	**Semicarbazone**
2-Methyl-propanal	-----	64^0	182^0	131^0	125^0
Butanal	----	75	123	91	106
But-2-enal	---	104	190	185	199
Hexanal	---	131	107	80	106
2-Furaldehyde	---	161	---	152	203
Benzaldehyde	---	179	237	192	224
Salicylaldehyde	---	197	252	228	231
4-Methyl-benzaldehyde	---	201	233	201	234
2-Chloro-benzaldehyde	---	213	209	249	---
3-Chloro-benzaldehyde	--	214	----	216	229
4-Methoxy-benzaldehyde	---	248	253	160	210
Cinnamaldehyde	--	248	255	195	215
3,4-Methylenedioxy-benzaldehyde	37^0	--	271	199	237
2-Methoxy-benzaldehyde (2-Anisaldehyde)	37	263	253	205	215
4-Chloro-benzaldehyde	47	---	265	219	230
4-Hydroxy-3-methoxy-benzaldehyde (Vanillin)	81	---	269	228	239

Table IX Phenols.

Melting points of derivatives					
Phenol	**M.p.**	**B.p.**	**Benzoate**	**4-Toluene sulphonate**	**Aryloxyacetic acid**
4-*t*-Butylphenol	99°	--	82°	109°	86°
2-Chlorophenol	---	175	---	74	145
2-Methoxy phenol	--	205	58	85	119
3-Chloro phenol	33	214	71	47	109
4-Chloro phenol	43	217	86	71	156
2,4- Dichlorophenol	45	210	96	125	140
Methyl salicylate	--	223	92	---	--
Ethyl salicylate	---	231	80	---	--
4-Hydroxy-3-methoxy-benzaldehyde	81	---	78	115	176
4-Chlor-2-methyl phenol	48	222	71	---	115
Ethyl-4-hydroxybenzoate	116	----	94	---	----
Methyl-4-hydroxybenzoate	131	----	135	-----	----

Table X Ketones.

Melting Points of Derivatives					
Ketone	**M.p.**	**B.p.**	**2,4-Dinitro-phenylhydazone**	**4-Nitro-Phenyl-hydrazone**	**Semi-Carbazone**
Acetone	--	56°	128°	148°	189°
Butanone	---	80	115	128	146
3-Methylbutane-2-one	---	94	120	109	114
Pentane-2-one	---	102	141	117	108
Pentane-3-one	---	102	156	144	139
4-Methylpentane-2-one	---	117	95	79	132
2,4-Dimethylpentane-3-one	---	124	94-98	---	160
Cyclopentanone	---	131	146	154	214(d)
Cyclohexanone	---	156	162	147	167
2,6-Dimethylheptane-4-one	---	168	92	---	122

Table 3.10 ***contd…***

Melting Points of Derivatives					
Ketone	M.p.	B.p.	2,4-Dinitro-phenylhydazone	4-Nitro-Phenyl-hydrazone	Semi-Carbazone
Acetophenone	---	202	240	185	199
Propiophenone	---	218	191	----	174
4-Methyl-acetophenone	--	224	258	198	205
4-Chloro-acetophenone	---	236	---	239	201
4-Methoxy-acetophenone	39^0	---	220	---	198
4-Phenyl-3-butene-2-one	42	262	227	166	186
Benzophenone	48	--	238	165	---

Table XI carboxylic acids.

Melting points of derivatives					
Acid	M.p.	B.p	Amide	Anilide	Anhydride
But-2-enoic acid	72^0	--	160^0	118^0	-------
Phenylacetic acid	76	--	157	118	-------
Phenoxyacetic acid	99	--	101	101	----
2-Methoxybenzoic acid	101	---	129	131	----
2-Methylbenzoic acid	105	---	143	125	-----
3-Methylbenzoic acid	111	----	95	126	-----
Benzoic acid	121	----	129	162	-----
Cinnamic acid	133	----	147	153	-----
2-Furoic acid	133	---	142	123	----
Maleic acid	135	---	---	---	56^0
2-Chlorobenzoic acid	141	----	141	118	------
3-Nitrobenzoic acid	141	---	142	154	-----
2-Nitrobenzoic acid	147	---	175	155	----
3-Chlorobenzoic acid	158	---	134	124	----
4-Methylbenzoic acid	178	---	159	146	----
4-Methoxybenzoic acid	184	---	162	169	----
Succinic acid	185	----	-----	----	120
3,5-Dinitrobenzoic acid	207	----	183	234	-----
4-Nitrobenzoic acid	239	---	201	211	------
4-Chlorobezoic acid	243	----	179	194	-----

Table XII Esters.

Melting points of derivatives			
Ester	**M.p**	**B.p**	**Carboxylic acid**
Methyl but-2-enoate	----	119^0	72^0
Ethyl but-2 enoate	----	138	72
Methyl benzoate	---	199	121
Ethyl benzoate	---	213	121
Diethyl maleate	-----	225	135
Ethyl phenylacetate	----	229	76
Ethyl cinnamate	----	271	133
Methyl cinnamate	36^0	----	133
Ethyl 4-nitrobenzoate	57	----	239
Phenyl benzoate	69	----	121
Methyl 3-nitrobenzoate	79	---	141
Methyl 4-nitrobenzoate	96	---	239
Methyl 3,5-dinitrobenzoate	107	---	207

Table XIII Anhydrides.

Melting points of derivatives					
Anhydride	**M.p**	**B.p**	**Carboxylic acid**	**Amide**	**Anilide**
Acetic anhydride	----	140^0	----	82^0	114^0
Propanoic anhydride	----	167	----	81	106
Maleic anhydride	56^0	----	130	181 (mono)	174 (mono)
Succinic anhydride	120	---	185	157 (mono)	180 (mono)
Phthalic anhydride	132	---	205	149 (mono)	170 (mono)

Table XIV Amides.

Melting points of derivatives			
Amide	**M.p**	**Carboxylic acid**	**Amine**
Benzamide	129^0	121^0	----
4-Ethoxyacetanilide (phenacetin)	135	----	b.p. 250^0
Benzanilide	163	121	----
4-Chloroacetanilide	179	---	m.p. 70^0
4-Methyl-2-nitroacetanilide	96	---	m.p. 116^0

Table XV Nitriles.

Melting points of derivatives		
Nitrile	**B.p**	**Carboxylic acid**
Benzonitrile	192^0	121^0
Benzyl cyanide	234	76

Table XVI Amines.

Amine	m.p	b.p	Acetamide	Benz-amide	Toluene-*p*-sulphonamide	2,4-dinitro-phenyl derivative	3,5-dinitro benzamide
2-Butylamine	--	63^0	-----	84^0	-----	---	162^0
Isobutyl-amine	--	68	----	---	78	80	158
1-Butyl-amine	---	77	---	---	---	90	102
Cyclohexyl-amine	---	134	---	149	87	156	207
Benzyl-amine	---	185	---	105	116	---	---
4-Ethoxy-aniline	---	250	137	173	106	118	---
Methyl-2-amino-benzoate	---	255	101	100	---	---	---
Ethyl-4-amono-benzoate	92^0	---	110	148	----	---	---
3-Amino-aceto-phenone	99	----	128	---	130	---	---
4-Methyl-2-nitro-aniline	116	---	93	144	---	---	--
2-Chloro—5-nitro-aniline	126	---	156	---	---	--	--
2-Methyl-4-nitro-aniline	130	---	202	174	---	--	--

3.8 Procedure for the Preparation of Derivatives

In this section of the manual you will find standard procedures for the preparation of all the various types derivatives listed in the Table VII-XVI. They are listed under each functional group, and these appear in the same order as the tables: Alcohol, Phenols, Aldehydes, Ketones, Carboxylic acids,

Esters, Anhydrides, Amides, Nitriles and Amines. Remember some compounds with more than one functional group may appear in more than one table. It is important that you know the molecular structure of each derivative that you prepare and understood the chemistry of its formation. Most of the unknown compounds are named systematically, but a few are listed using their unsymmetrical (or trivial) names. In these cases, if you are unsure what the molecular structure is, you should consult either the "Aldrich" catalogue or The Dictionary of Organic Compounds".

You should determine the best possible solvent for recrystallisation for each derivative, using a small sample, taking care not to draw it in too much solvent. Occasionally, you will have to do a recrystallisation using a mixed solvent system, e.g. dichloromethane and petroleum ether (60-80^0), this depends on the differing solubilities of the solid you are attempting to recrystallise in the two solvents. The idea here is to dissolve the solid in the solvent in which it is most soluble and while the solution is still hot, add the other solvent using a Pasteur pipette until the warm solution goes hazy. When allowed to cool, the solid usually crystallizes out, but the success of this type of recrystallisation depends on using the absolute minimum of the first solvent.

Note: Remember when crystallizing from water that it will not be sufficient to use a steam bath, to get the water boiling you must use either a stirrer hot plate or a Bunsen flame as your heat source.

(Care! Glassware will crack if heated directly with a naked flame).

3.8.1 Alcohols

(i) Preparation of 3,5-dinitrobenzoate (Ensure all glassware for this procedure is dry)

Dissolve 3,5-dinitrobenzoyl chloride (0.3 g) in pyridine (2 ml) in a 10 ml flask equipped with a reflux condenser and a calcium chloride drying tube. Dissolve the alcohol (0.3-0.5 ml) in pyridine (2 ml) and add to the acid chloride solution. Warm the mixture on a steam bath for a minimum of 5 min. (primary alcohol) 10 min. (secondary alcohol); 30 min. (tertiary alcohol).

N.B. The 3,5-dinitrobenzoyl chloride is stored in pre-weighed vials in a desiccator in the fume cupboard. Use the contents of one vial per attempt, there is no need to reweigh.

Pour the cooled reaction mixture into water (50 ml) and extract with dichloromethane (25 ml). Wash the separated dichloromethane layer with dil. HCl (25 ml x 2) and then with sodium bicarbonate solution (25 ml). Dry over Na_2SO_4, filter and evaporate off the solvent using the rotary evaporator. Crystallize the residue from a suitable solvent (Petroleum ether 60-80^0 C).

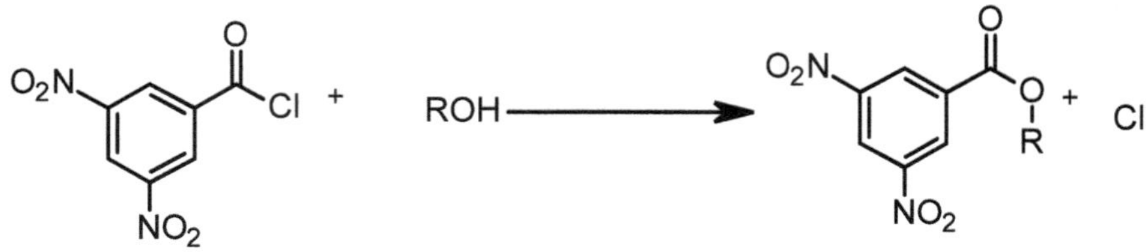

3.8.2 Preparation of 4-nitrobenzoate (Ensure all glassware for this procedure is dry)

(a) Preparation of the reagent 4-nitrobenzoyl chloride: Place equal quantities of 4-nitrobenzoic acid and phosphorous pentachloride (approx. 1 g each) in a test tube and warm gently using a small Bunsen flame in the fume cupboard.

The contents of the tube will melt and vigorous evolution of HCl will occur. Continue heating by passing in and out of the Bunsen flame just often enough to keep the reaction going, until evolution of HCL stops. Pour the reaction mixture into a watch glass, place in a desiccator and allow to cool. The crude acid chloride will solidify, remove the phosphorous oxychloride, which is formed in the reaction, by pressing the crude solid on a porcelain plate. (Please wash the piece of plate before putting in the bin). The 4-nitrobenzoyl chloride is pure enough to use for the next stage.

(b) Now prepare the **4-nitro benzoate ester** using the same procedure described above for the 3,5-dinitrobenzoate ester. Crystallise from a suitable solvent (petroleum ether 60-80^{0}C).

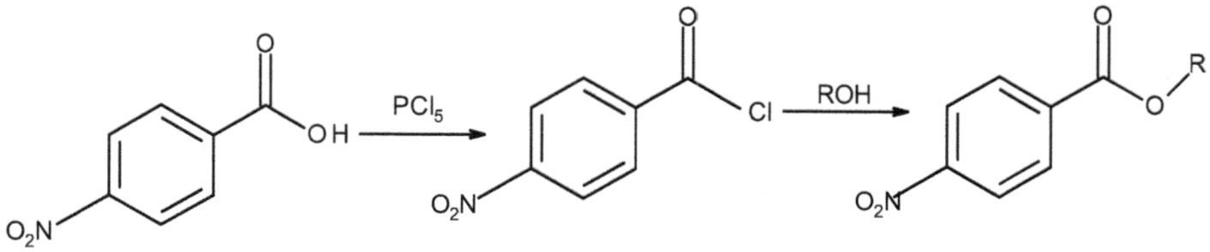

3.8.3 Preparation of Hydrogen 3-nitrophthlate

(Ensure all glass ware for this procedure are dry)

3-Nitrophthalic anhydride is stored in a desiccator in the fume cupboard. It is already pre-weighed. Use the contents of one vial. Heat the reagent (0.5g) with the alcohol (2 ml) on a steam bath for 30 min. While still hot pour into water (10 ml), stir well and leave it to stand until a solid separates. In some cases you may have to leave it standing until the next laboratory period. Filter off the solid and dry in a vacuum desiccator over phosphorus pentoxide (leave for 1 hour). Recrystallise from a mixture of dichloromethane and petroleum ether 60-80 °C.

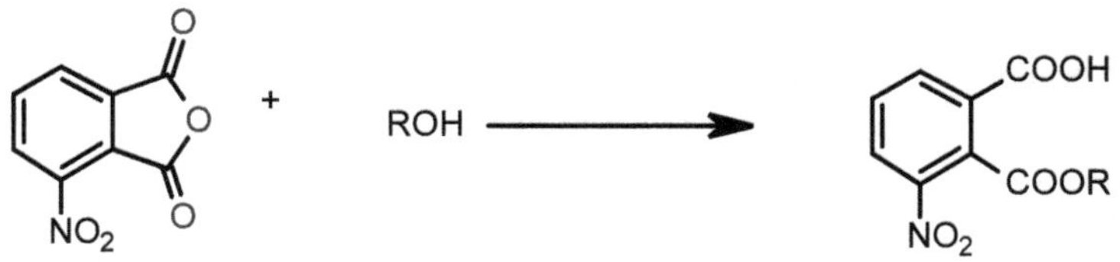

N.B. Can you explain why the carbonyl next to the nitro group is more reactive.

3.9 Phenols (wear gloves when handling phenols)

3.9.1 Preparation of Benzoate Derivative

(Measure out the benzoyl chloride in a fume cupboard. Take care not to exceed the amount given)

Place dil. sodium hydroxide solution (7 ml) in a measuring cylinder and dilute to 25 ml with water. Use this reagent to dissolve the phenol (1 g) in 50 ml round bottom flask. Add benzoyl chloride (1 ml). Stopper the flask and shake vigorously for 10-20 minutes. Check the solution is still alkaline, if not, add more of the NaOH solution (1 ml). If the derivative fails to solidify at first, cool in an ice bath (if sample is still an oil leave to stand overnight. Filter off the precipitate and crystallize from a suitable solvent (methanol, ethanol).

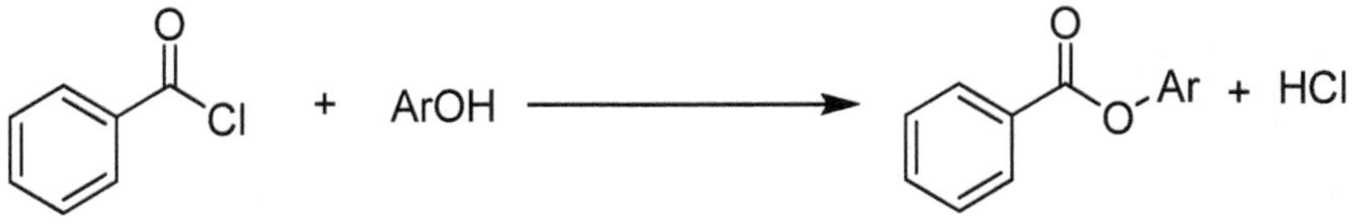

3.9.2 Preparation of 4-Toluene Sulphonate

(Ensure all glass ware for this procedure is dry).

Warm a mixture of the phenol (1 g), 4-toluenesulphonyl chloride (1.5 g) and pyridine (3 ml) in a 10 ml round bottom flask equipped with a reflux condenser and a calcium chloride drying tube on a steam bath for 15 min. Cool and pour the mixture onto a mixture of ice and conc. HCl. Continue cooling until the derivative crystallizes out. Recrystallise from a suitable solvent (ethanol, methanol).

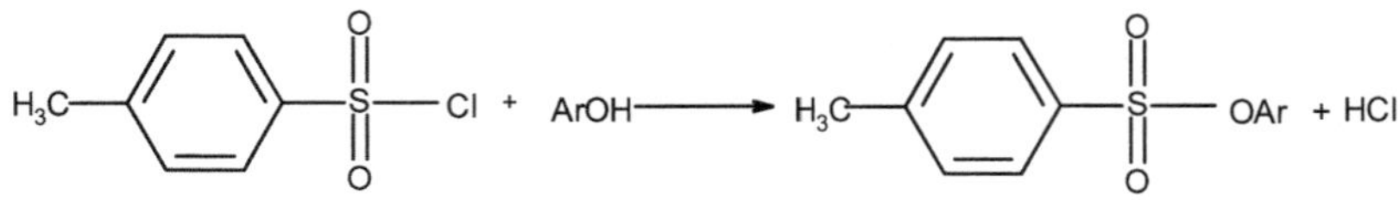

3.9.3 Preparation of Aryloxyacetic Acid

Dissolve or suspend the phenol (1 g) in conc. NaOH (5 ml) in a test tube, and add 505 aqueous chloroacetic acid (4 ml). Ensure the reagents are well mixed and warm the test tube in the steam bath for 15 min. It's easiest just to stand the test tube in the steam bath, the solution should go clear (In the case of t-butyl phenol, it goes clear then a solid precipitates out). Cool and add water (10 ml). Acidify with conc. HCl, cool and filter off the precipitate. Recrystallise from a suitable solvent (Aqueous ethanol, water).

3.10 Aldehydes and Ketones

3.10.1 Preparation of 2,4-Dinitrophenylhydrazone

Add the aldehyde or ketone (0.4 g) {N.B. If the unknown is a solid dissolve it in methanol (2-3 ml) first} in Brady's reagent (25 ml). Warm on a steam bath for 5 min. Make sure the reaction mixture is boiling. Cool and filter off the orange red precipitate. Wash well with lots of cold water. Recrystallise from a suitable solvent (ethanol, methanol mixture of dichloromethane and petroleum ether (60-80^0 C).

3.10.2 Preparation of 4-Nitrophenylhydrazone

Dissolve 4-nitrohydrazine (0.3 g) in methanol (2 ml) and add glacial acetic acid (one drop). To this solution add a solution of the unknown sample (0.2 g) in methanol (2 ml), heat the mixture on the steam bath for 5 min. Cool, if no precipitate forms and add a few drops of water until the solution just becomes cloudy. Filter off the precipitate and recrystallise from methanol.

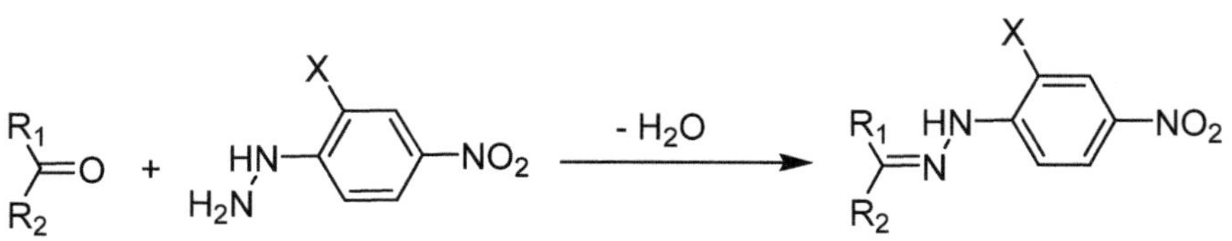

Ketones: R^1, R^2 = alkyl or aryl;

Aldehydes: R^1 = alkyl or aryl; R^2 = H

2,4-Dinitrophenylhydrazones: X = NO_2

4-Nitrophenylhydrazones: X = H

3.10.3 Preparation of Semicarbazone

In a small flask place semicarbazide hydrochloride (0.3 g) and sodium acetate trihydrate (0.5 g). Add a few drops of water to dissolve the mixture (gentle warming may be required). Dissolve the unknown sample (0.2 g) in methanol (2 ml) and add to the reagent. If no precipitate forms warm on a steam bath for 5 min.. Cool and filter off the semicarbazone. Recrystallise from a suitable solvent (methanol, ethanol).

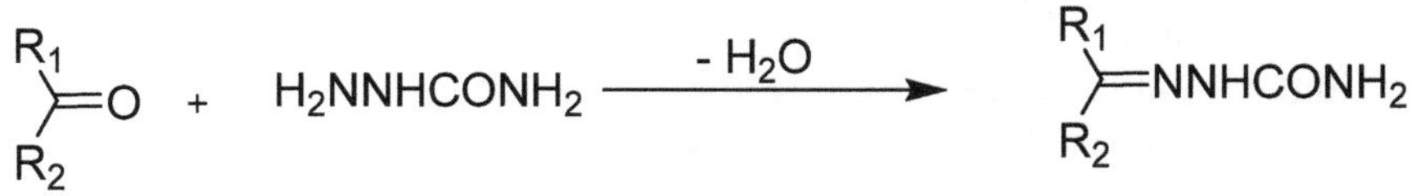

Ketones: R^1, R^2 = alkyl or aryl; Aldehydes: R^1 = alkyl, aryl; R^2 = H.

3.11 Carboxylic Acids

3.11.1 Preparation of Amides and Anilides

These are made by the reaction of the acid chloride with ammonia or aniline

(Ensure all glassware for this procedure is dry and carryout reaction in a fume cupboard).

(a) ***First prepare the acid chloride*****:** (except for 3, 5-dinitrobenzoic, 4-nitrobenzoic or 4-chlorobenzoic acids this method for the preparation of the acid chloride will not work).

Heat a mixture of carboxylic acid (1 g) with thionyl chloride (5 ml) in a 10 ml round bottom flask equipped with reflux condenser and a calcium chloride drying tube on a steam bath for 10 min. All the acid chloride dissolve into solution. Turn off the steam and set up the apparatus to distil off the excess thionyl chloride. Reheat to begin the distillation, after about 10 min. add 25 ml of petroleum ether (40-60^{0}C) and distil off the petrol. This should remove the last traces of thionyl chloride from the acid chloride, which, in most cases, will be a liquid. Reset the apparatus to reflux configuration and cool the acid chloride in some ice water. Use this to make either the amide, or anilide by one of the procedures (b) or (c) described below.

$$R-CO_2H+SOCl_2 \rightarrow R-COCl+SO_2+HCl$$

Preparation of the acid chloride for 3,5 dinitrobenzoic, 4-nitrobenzioc and 4-chloride benzoic acids:

Place equal quantities of the acids and phosphorous pentachloride (approx. 1g of each) in a test tube and warm gently using a small Bunsen flame in the fume cupboard. The contents of the tube will melt and vigorous evolution of HCl will occur. Continue heating by passing in and out of the Bunsen flame just often enough to keep the reaction going, until evolution of HCl stops. Pour the reaction mixture on to a watch glass, place in a desiccator and allow to cool. The crude acid chloride will solidify, remove the phosphorous oxychloride, which is formed in the reaction, by pressing the crude solid on a porcelain plate. Place the acid chloride in a small round bottom flask fitted with a reflux condenser and cool in ice water. Use this to make either amide or anilide by one of the procedures (b) or (c) described below.

$$R-CO_2H+PCl_5 \rightarrow R-COCl+POCl_3+HCl$$

(b) ***Preparation of an amide*:** To the cooled acid chloride carefully add conc. ammonia (5 ml) down the side of the condenser. Filter off the amide and wash well with water. If a solid does not precipitate on cooling, extract this aqueous solution with dichloromethane (2 ml x 2), combine the dichloromethane extracts and wash with dilHCl (2 ml x 2) to remove the excess ammonia. Dry the dichloromethane over Na_2SO_4, filter and evaporate to give a solid residue which should be recrystallised using a suitable solvent (ethanol, methanol, aqueous ethanol, aqueous methanol and occasionally water).

$$R-COCl+2NH_3 \rightarrow R-CONH_2+NH_4^++Cl^-$$

(c) ***Preparation of an anilide*:** To the cooled acid chloride add aniline (1 ml) drop wise. Cool the mixture if a precipitate is not formed immediately. Filter off the derivative and wash well with dil. HCl. If a solid does not precipitate on cooling, add some water (5 ml) and extract this aqueous solution with dichloromethane (2 ml × 2), combine the dichloromethane extracts and wash with dil. HCl (2 ml × 2) to remove the excess aniline. Dry the dichloromethane over Na_2SO_4, filter and evaporate to give a solid residue which should be recrystallised using a suitable solvent (ethanol, methanol, aqueous ethanol).

$$R\ COCl+2\ PhNH_2 \rightarrow R\ CONHPh+PhNH_3+Cl^-$$

3.11.2 Preparation of an Acid Anhydride
(Ensure all glassware for this procedure is dry)

Some carboxylic acids are easily dehydrated to form an intra-molecular (cyclic) acid anhydride. This is especially true where the ring formed would contain 5 or 6 atoms.

Reflux the acid (1 g) with acetic anhydride (1 ml) in a 10 ml round bottom flask equipped with a reflux condenser and a calcium chloride drying tube for 30 min. (oil bath and stirrer hot plate required). Cool in an ice-salt bath. Filter off the anhydride, wash the solid with a little ice-cold dry ether and quickly suck dry. Crystallize from a suitable solvent (ether, dichloromethane).

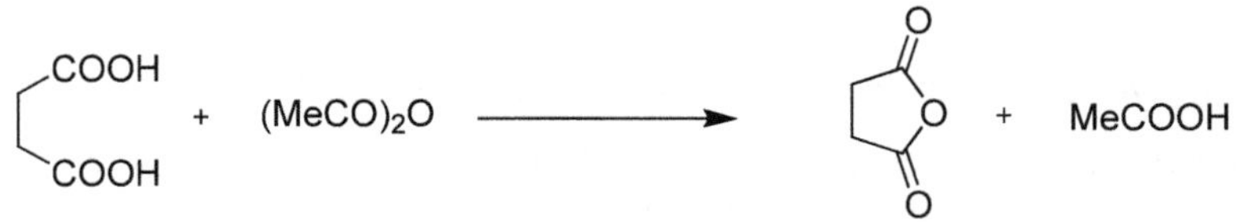

3.12 Esters

Hydrolysis of esters

It is usually necessary to hydrolyze the ester to the corresponding carboxylic acids and alcohols/phenols to obtain satisfactory derivatives. In most instances it is sufficient to isolate the carboxylic acid.

Make 10% ethanolic KOH by dissolving KOH (1 g) in ethanol (10 ml). Reflux the ester (2 g) with this KOH solution in a 25 ml round bottomed flask equipped with a reflux condenser for 30 min. (oil bath and stirrer hotplate required). Cool and dilute with cold water (25 ml). Wash with dichloromethane (10 ml x 2) to remove any un-reacted ester that may remain. Acidify the aqueous layer with conc. HCl and cool in ice. If the carboxylic acid precipitates out filter it off, and wash with a little cold water. Take care, some carboxylic acids are surprisingly soluble in water! To maximize the yield of carboxylic acid you should extract the aqueous filtrate with dichloromethane (15 ml x 2). Wash the combined dichloromethane extracts with water, dry over sodium sulphate, filter and evaporate using a rotary evaporator. If the carboxylic acid does not precipitate on acidification, then extract the aqueous solution with dichloromethane as described above for the filtrate. Combine all solids and crystallise from a suitable solvent. (petroleum ether (60-80^0 C) or dichloromethane, water).

$$R^1CO_2R^2 + KOH \rightarrow R^2OH + R^1CO_2K \rightarrow R^1CO_2H + KCl$$

3.13 Anhydrides

3.13.1 Hydrolysis to Carboxylic Acid

The anhydride can be hydrolyzed by warming with water or dilute NaOH. Cool the resulting solution, acidify with conc. HCl and filter off the solid acid. Crystallize from a suitable solvent (dichloromethane, water).

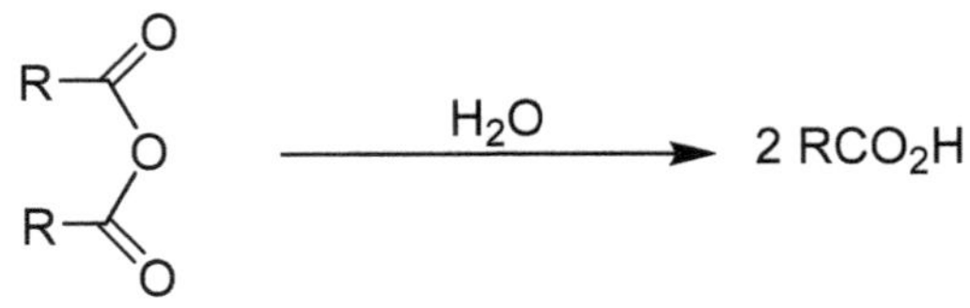

N.B. A cyclic anhydride will hydrolyze to give a dicarboxylic acid.

Preparation of amides and anilides: Amides and anilides can be prepared directly from anhydrides using procedures similar to those described starting from acid chlorides.

3.13.2 Preparation of an Amide

To the cooled anhydride CAREFULLY add conc. ammonia (5 ml) down the side of the condenser. Filter off the amide, and wash well with water.

3.13.3 Preparation of an Anilide

To the cooled acid anhydride add aniline (1 ml) drop wise. Cool the mixture if a precipitate is not formed immediately. Filter off the derivative and wash well with dilute hydrochloric acid.

In both the cases, if a solid does not precipitate on cooling, you will have to do an extraction. In the case of an anilide add some water (5 ml), to give you an aqueous solution to extract, there should be enough water already present in the ammonia used to make the amide so DO NOT add any more. Extract this aqueous solution with dichloromethane (2 ml x 2), combine the dichloromethane extracts and wash with dil. HCl (2 ml x 2) to remove the excess ammonia/aniline. Dry the dichloromethane over sodium sulphate, filter and evaporate to (hopefully) give a solid residue which should be recrystallised using a suitable solvent (for amides: (ethanol, methanol, or aqueous ethanol/methanol and occasionally water) for anilides: (ethanol, methanol, aqueous ethanol).

$$R-CO-O-R+R^1NH_2 \rightarrow R-CO-NH-R^1+RCO_2H$$

$$R^1 = H \quad \text{(amides)}$$

$R^1 = Ph$ (anilides)

N.B. A cyclic anhydride will react to give a mono acid/mono amide or mono acid or mono anilide.

3.14 Amides

3.14.1 Hydrolysis of Primary Amides to Carboxylic Acid

Reflux the amide (1 g) with conc. sodium hydroxide solution (10 ml) in a 25 ml round bottomed flask equipped with a reflux condenser for about 30 minutes (oil bath and stirrer hotplate required) until no more ammonia is evolved (test at top of condenser with pH paper). Cool and dilute with cold water (25 ml). Wash with dichloromethane (10 ml x 2) to remove any unreacted amide that may remain. Acidify the aqueous layer with con. HCl and cool in ice. If the carboxylic acid precipitates out filter it off, and wash with a little cold water. Take care, some carboxylic acids are surprisingly soluble in water! To maximize the yield of carboxylic acid you should extract the aqueous filtrate with dichloromethane (15 ml × 2). Wash the combined dichloromethane extracts with water, dry over Na_2SO_4, filter and evaporate using a rotary evaporator. If the carboxylic acid does not precipitate on acidification, then extract the aqueous solution with dichloromethane as described above for the filtrate. Combine all solids and crystallise from a suitable solvent (petroleum ether (60-80^{0}C), dichloromethane, water).

$$R\ CONH_2 + NaOH \rightarrow R\ CO_2Na^+ + NH_3 \rightarrow R\ CO_2H$$

3.14.2 Hydrolysis of N-Substituted Amides

These are difficult compounds to identify as they must be hydrolyzed to the constituent acid and amine. It is usually enough to characterize only one of the two constituents, usually the one which is solid (in most cases this will be an aromatic compound). Thus if the amide is formed from an aromatic amine (i.e. an anilide) and an aliphatic acid it is usually best to isolate and identify the amine.

Make 67% H_2SO_4 by CAREFUL addition of conc. H_2SO_4 (5 ml) to water (5 ml). Reflux the amide (2 g) with this acid in a 25 ml round bottom flask equipped with a reflux condenser for 1 hr. (oil bath and stirrer hotplate required). Dilute the solution with water (10 ml) and cool. The solution will contain the free acid and the sulphate salt of the amine. The acid may crystallize out on cooling in ice. If so, filter off and wash well with cold water. if no solid forms extract the solution with dichloromethane (25 ml x 2). Wash the combined extracts with water, dry over Na_2SO_4, filer and evaporate using a rotary evaporator. Crystallize the acid form with dichloromethane/water. After removal of the acid constituent, make the solution strongly basic by adding conc. sodium hydroxide solution. Extract the mixture with dichloromethane (25 ml x 2), wash the combined extracts with water, dry over Na_2SO_4, filter and evaporate using a rotary evaporator. If the amine is solid crystallise from ethanol/water, if not then a suitable derivative of the amine must be made.

$$R^1CONHR^2 + H_2SO_4 \rightarrow R^1CO_2H + (R^2NH_3)_2SO_4$$

$$(R^2NH_3)_2SO_4 + 2NaOH \rightarrow 2R^2NH_2 + Na_2SO_4 + 2H_2O$$

3.15 Nitriles

3.15.1 Hydrolysis to Carboxylic Acid

Nitriles are characterized by hydrolysis to the parent carboxylic acid.

Make 67% H_2SO_4 by CAREFUL addition of conc. H_2SO_4 (5 ml) to water (5 ml). Reflux the nitrile (2 g) with this acid in a 25 ml round bottom flask equipped with a reflux condenser for 1 hr. (oil bath and stirrer hot plate required). Dilute the solution with water (10 ml) and cool. The solution will contain the free acid. The acid may crystallize out on cooling in ice. If so, filter off and wash well with cold water. If no solid forms extract the solution with dichloromethane (25 ml × 2). Wash the combined extracts with water, dry over Na_2SO_4, filter and evaporate using a rotary evaporator. Crystallise the acid from a suitable solvent (petroleum ether (60-80^0 C), dichloromethane or water).

$$RCN + H_3O^+ + H_2O \rightarrow RCO_2H + NH_4^+$$

3.16 Amines

(WEAR GLOVES WHEN HANDLING AMINES)

3.16.1 Preparation of Acetamides

Add acetic anhydride (3 ml) to the amine (0.5 g). The reaction mixture should become hot and the colors of strongly colored amine should diminish. If this does not happen add a drop off conc. H_2SO_4 to the reaction mixture. Leave the reaction to stand for 15 min. and then pour the mixture into a beaker of cold water (25 ml). Filter off the solid and recrystallise from a suitable solvent (methanol, ethanol, aqueous ethanol, water).

$$RNH_2 + Ac_2O \rightarrow RNHCOCH_3 + CH_3CO_2H$$

R = H: Amides

R = Ph: Anilides

3.16.2 Preparation of Benzamide Derivative (Carry out reaction in a fume cupboard)

Make 1M solution of sodium hydroxide by adding water (5 ml) to dil. sodium hydroxide solution (5 ml) in a 25 ml round bottom flask. Suspend the amine (0.5 g) in this solution and add benzoyl chloride (1 ml). Stopper the flask and shake vigorously for 15 min. Cool the reaction mixture and check it is still alkaline using pH paper. If not add some more 1M sodium hydroxide solution. The derivative should solidify, if not cool in an ice salt bath with intermittent shaking until solid forms. Filter of the derivative, wash well with water and crystallise form a suitable solvent {petroleum ether (60-80 °C), ether, ethanol methanol}.

3.16.3 Preparation of 3,5-Dinitrobenzamide

Dissolve 3,5-dinitrobenzoyl chloride (0.3 g) in pyridine (2 ml) in 10 ml flask equipped with a reflux condenser and a calcium chloride drying tube. Add the amine (0.1 g) to the acid chloride solution. Warm the mixture on a steam bath for 15 min. Pour the cooled reaction mixture into water (50 ml) and extract with dichloromethane (25 ml). Wash the separated dichloromethane layer with dil. HCl (25 ml × 2) and then with sodium bicarbonate solution (25 ml). Dry over Na_2SO_4, filter and evaporate off the solvent using the rotary evaporator. Crystallise the residue from mixture a of dichloromethane and petroleum ether (60-80^{0}C).

N.B. The 3, 5-dinitrobenzoyl chloride is stored in pre-weighed vials in a desiccator in the fume cupboard. Use the contents of one vial per attempt, there is no need to reweigh.

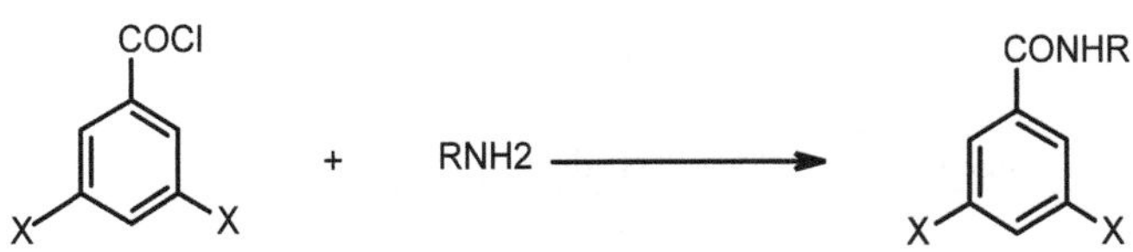

Benzamide X = H

3,5-dinitrobenzamide X = NO_2

3.16.4 Preparation of Toluene-4-Sulphonamide

Use the same procedure as per the benzamide but use toluene-4-suphonyl chloride (1.5 g) instead of benzyl chloride. Crystallise from a suitable solvent {petroleum ether (60-80^{0}C), ethanol, methanol}.

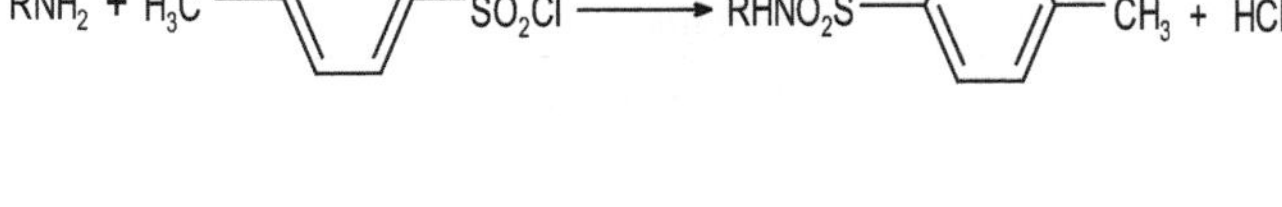

3.16.5 Preparation of 2,4-Dinitrophenyl Derivatives

Dissolve the amine (0.5 g) and 2,4-dinitrochlorobenzene in the minimum amount of boiling ethanol. Reflux (condenser required) for 30 min. cool in an ice bath, filter off precipitate and crystallise from ethanol.

Caution: 2,4-dinitrochlorobenzene is a potent skin irritant. Gloves must be worn at all times when making 2,4-dinitrophenyl derivatives of amines.

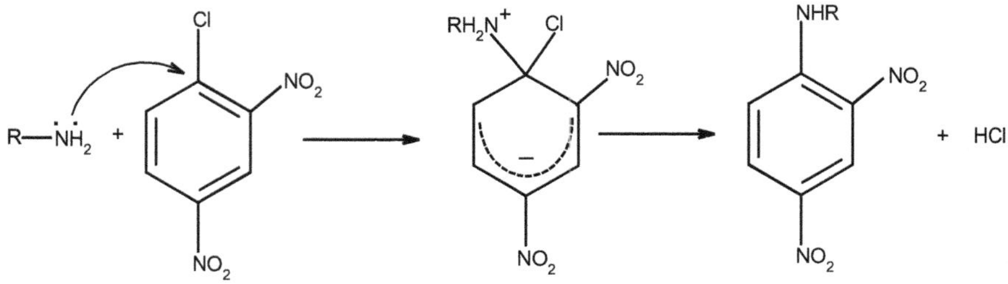

3.17 Specimen Report

The identification of a single organic substance containing nitrogen.

Test	Observation	Inference
Appearance and smell	Pale yellow Crystalline solid Faint smell of almonds	Possibly unsaturated
Melting point	57-58 ^{o}C	
Solubility: H_2O pH dil. HCl dil. NaOH	Slightly soluble hot Slightly < 7 Insoluble Insoluble	Fairly polar compound Neutral? Not amine Not acid or phenol
Heated with dil. NaOH	No NH_3 detected	Not primary amide
Unsaturation: Br_2/CCl_4 Br_2/H_2O	No reaction evident No reaction	No alkene C=C or easily brominated compound
$KMnO_4$	Decolorisation after a few seconds	Easily oxidized group present
Carbonyl tests: Brady's reagent (2,4-dinitrophenylhydrazine) Tollen's test ($AgNO_3/NH_4OH$)	Yellow-orange precipitate Silver mirror developed slowly	>C=O present (aldehyde or ketone) -CHO group present
$TiCl_3$ solution	Solution decolorized on boiling with unknown compound	$-NO_2$ group present
$FeCl_3$ solution	No color change	No phenol or enol

Inference: The compound is a nitro-aldehyde (from tables it could be 3-nitrobenzaldehyde (lit. m.p. 58 ^{O}C)

3.18 Infrared Spectrum (Nujol mull)

Interpret infrared spectra as fully as possible. (Do not take Nujol bands as belonging to your unknown).

3060 cm^{-1} (w) - Aromatic ν(CH)

1705 cm^{-1} }
- ν(C=O) consistent with aromatic aldehyde
1685 cm^{-1}}

1615 cm^{-1} (m) Aromatic skeletal vibration

1530 cm^{-1} (s) and 1340 cm^{-1} ν_{asym} of NO_2 group (ν_{sym} likely to be obscured by Nujol band at 1340 cm^{-1}).

805 cm^{-1} (m) and 725 cm^{-1} (s) 1,3-Disubstituted benzene ring

Inference: The compound is an aromatic carbonyl compound possibly 1,3 disubstituted; spectrum consistent with presence of both an aromatic aldehyde and a nitro group.

3.19 ^{1}H NMR Spectrum

δH 10.2 singlet -C**H**O (aldehyde)

δH 8.7 singlet

δH 8.5 and δH 8.25 doublets with J = 5 – 7 Hz - *meta* coupling

δ7.8 apparent triplet J = 6 Hz - *meta* coupling

Integrals 1 (aldehyde) : 4 (aromatic)

Inference: Not 1,4- disubstituted; aromatic singlet suggests 1,3 - disubstitution.

Predicted chemical shifts in position of benzene protons (δH 7.27) caused by substituents (from tables)

Substituent	*ortho*	*meta*	*para*
-CHO	+ 0.56	+ 0.22	+ 0.29
$-NO_2$	+ 0.95	+ 0.26	+ 0.38

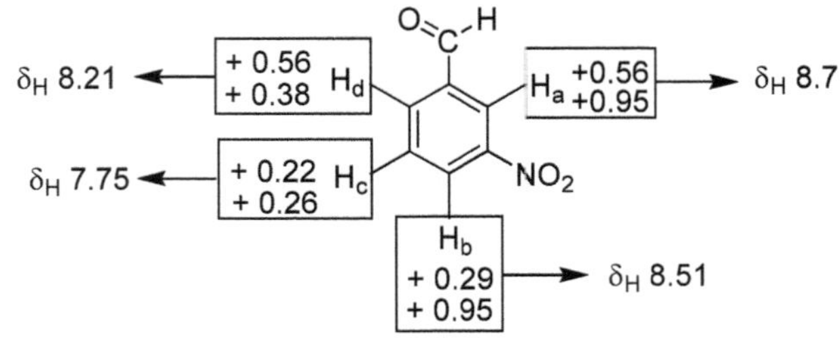

Assign all ^{1}H NMR bands. (So far as possible, account for all chemical shift values, integrations, and splitting patterns).

3.19.1 δ_H Values

^{1}H	Found	Predicted
CHO	10.2 (s)	-10
H_a	8.7 (s)	8.78 (s)
H_b	8.5 (d)	8.51 (d)
H_c	7.8 (t)	7.75 (t)
H_d	8.25 (d)	8.21 (d)

Inference: The predicted proton chemical shifts for 3-nitrobenzaldehyde are in good agreement with the observed values.

3.19.2 ^{13}C NMR Spectrum

Predicted influence of functional group substituents on chemical shift positions of benzene carbons (δ_C 128) (from tables).

Substituent	*ipso*	*ortho*	*meta*	*para*
-CHO	+ 9	+ 1	+ 1	+ 6
$-NO_2$	+ 20	- 5	+ 1	+ 6

Assign all ^{13}C NMR bands. (Make sure your assignments are consistent with the observed splittings by attached protons)

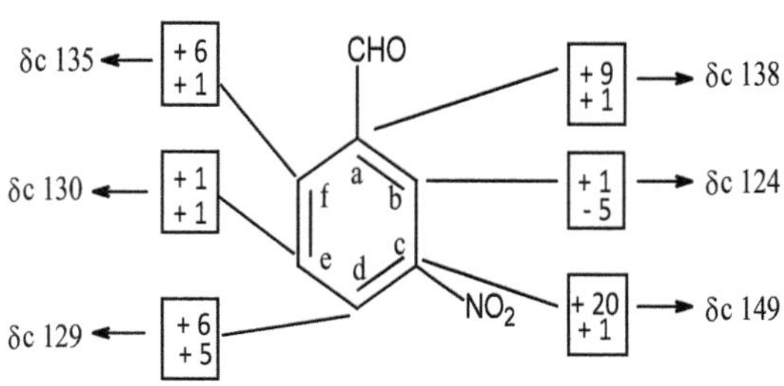

3.19.3 δ_c Values

	Found	Predicted
CHO	190.4 (d) – CH	190 – 194 (d)
C_a	137.9 (s) – quaternary	138 (s)
C_b	^{13}C	124 (d)
C_c	145.2 (s) – quaternary	149 (s)
C_d	128.9 (d) – CH	129 (d)
C_e	130.9 (d) – CH	130 (d)
C_f	135.2 (d) – CH	135 (d)

Inference: The predicted ^{13}C chemical shifts for 3-nitrobenzaldehyde are in good agreement with the observed values.

From chemical and spectroscopic data, the unknown compound is probably 3-nirobenzaldehyde.

3.20 Preparation of Derivative

1. 2,4-Dinitrophenylhydrazone, m.p. 280-285 ^{O}C

Recrystallised from ethanol /CH_2CL_2, m.p. 288-290^0 (Lit. m.p. 292 ^{O}C)

You should make only one derivative of each of unknown sample, but more than one attempt sometimes may be needed.

Always show its structure and the reaction in which it was formed.

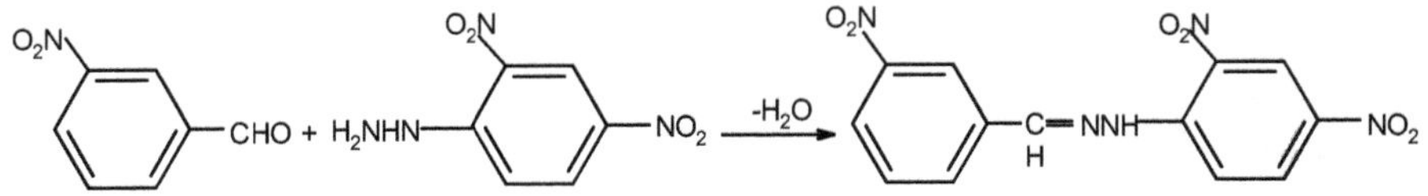

Conclusion: The unknown substance is 3-nitrobenzaldehyde

CHAPTER 4

Preparation of Organic Compounds

4.1 Calculation of Percentage Yield

In any preparative experiment you will be asked to calculate the yield. This requires that you first write a balanced equation for the synthesis

e.g: $ROH + R^1COOH \rightarrow R^1CO - OR + H_2O$

Then identify the minority reactant i.e., this example requires equal numbers of moles of alcohol and acid. Let us suppose that your calculations, from the masses of each used and their molecular weights, show that they are fewer moles of alcohol than acid i.e., the alcohol is the minority reactant. Thus we base our yield on this reactant (ROH).

Calculate moles ROH taken = mass (ROH) / M.W.(ROH)

Calculate moles of product = mass (R^1 CO-OR) / M.W. (R^1 CO-OR)

Then yield = {moles product / moles ROH X 100/1} %

Since only 1 mole of ester is produced from 1 mole of alcohol.

Example 1

For the preparation of n-butyl bromide the overall reaction is

$$C_4H_9OH + Br^- + H_2SO_4 \rightarrow C_4H_9Br + OH^-$$

This indicates that 1 mole of n-butanol reacts with 1 gm ion of bromide (as 1 mole of NaBr).

Inspection of the experimental details indicates that 0.17 mole of NaBr is required but only 0.135 mole of n-butanol is consumed. It is normal practice to calculate the percentage of a theoretical yield based upon the reactant present in smallest molar amount i.e., in this case the yield would be based upon n- butanol.

If this particular experiment gave a 100% yield of crude product, from 10 g. (0.135 mole) of n-butanol one would expect 0.135 mole of n-butyl bromide.

Since the molecular weight of n C_4H_9Br is 137

$$0.135 \text{ mole} = 0.135 \times 137 \text{ g} = 18.5 \text{ g.}$$

That is theoretically 10 g. $C_4H_9OH \rightarrow$ 18.5 g. C_4H_9Br = 100% yield.

In fact the experimental yielded only 12.4 g. C_4H_9Br

The actual yield is therefore 12.4/18.5 × 100 = 67% of the theoretical.

Example 2

Preparation of 4-nitrobenzyl alcohol, the overall reaction is

$$4\,NO_2ArCHO + NaBH_4 \rightarrow (ArCH_2O)_4B^-Na^+ + H_3O^+ \rightarrow 4\,NO_2ArCH_2OH$$

Which indicates that 4 moles of aldehyde are reduced by 1 mole of borohydride.

Inspection of the experimental details show 8 m mole of borohydride are used but only about 0.8 m mole is required. The borohydride is therefore present in large excess. The percentage yield of a product is normally based upon the reactant present in the smallest equivalent amounts, i.e. in this case the yield would be based on 4-nitrobenzaldehyde.

If this experiment gave a 100% yield of crude product, then one would obtain 3.3 m moles of the alcohol from 3.3 m moles of the aldehyde.

M.W. of 4-$O_2NC_6H_4CH_2OH$ = 153

i.e. 3.3 m moles = (3.3/1000) × 153 g = 0.507 g

Suppose that your experiment yielded only 0.36 g of the alcohol. Then

Actual yield (%) = 0.36/0.507 × 100 = 71

N.B. Quote yields only to the nearest whole percent – the accuracy of your weighings of reactants and product, and the purity of the reagents themselves cannot justify quoting the yield to the nearest 0.1%.

Preparative Organic Chemistry

This section of the book contains experiments from preparative organic chemistry.

The experiments listed below are essential before going to the next level of practical organic chemistry.

Important experiments

Preparation of an alkyl halide by an S_N1 reaction

4.2 2-Chloro-2-Methylpropane

Object: To convert 2-methylpropane-2-ol (t-butanol) into 2-chloro-2-methylpropane (t-butyl chloride) and to purify the product by distillation.

Background: Tertiary alcohols react very readily with concentrated hydrochloric acid to produce the corresponding chloride. With 2-methylpropane -2-ol the reaction is:

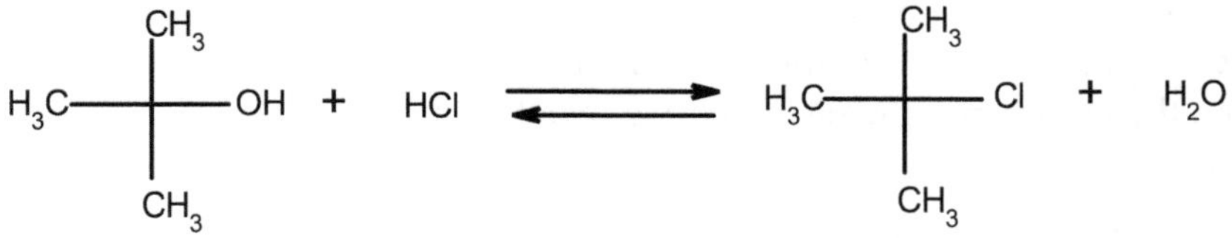

The mechanism is of the S_N1 type because the rate determining step involves the formation of an intermediate carbocation. A competing E_1 reaction, involving loss of a proton, yields a small amount of volatile 2-methylpropane (isobutene) as a by-product.

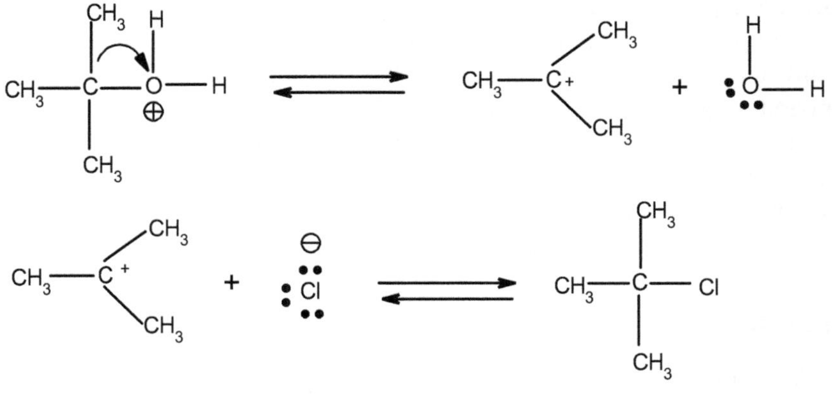

Experimental Procedure: Obtain 2-methylpropane-2-ol (0.34 mole) by either decanting 31.5 ml into a clean measuring cylinder or weighing 25 g into a clean beaker with the aid of a top-pan balance.

Take a 250 ml separating funnel and ensure that the tap is greased and free to turn. Close the tap and then add the alcohol, followed by concentrated hydrochloric acid (85 ml).

Stopper the funnel and shake the mixture periodically over 20 minutes, taking care to relieve any internal pressure build-up by inverting the funnel and opening the tap from time to time.

After 20 minutes, allow the funnel to stand for a few more minutes. Two immiscible layers will appear. The lower layer is *aqueous* and acidic.

Remove the stopper from the funnel, run off the lower layer through the tap into a beaker, and then dispose of it down the sink with plenty of running water.

The liquid remaining in the funnel is the impure 2-chloro-2-methylpropane. Add saturated sodium carbonate solution (10 ml) to the funnel and cautiously swirl the mixture several times before replacing the stopper. Then shake carefully, occasionally relieving any pressure build-up in the funnel.

Allow the mixture to stand for few minutes, then run off and discard the lower *aqueous* layer as before. Decant the impure 2-chloro-2-methylpropane from the funnel into a clean and dry conical flask.

Remove traces of water by adding anhydrous calcium chloride in successive small quantities. Filter of the solid residue using a clean, dry conical funnel and fluted filter paper. Collect the filtrate in a clean, dry pre-weighed 100 ml round –bottomed flask. Reweigh the flask and liquid to obtain a yield of crude product.

Set up a distillation apparatus over a steam bath, using clean and dry jointed equipment as shown in the diagram below, and distil the 2-chloro-2-methylpropane, collecting the fraction with b.p. about 51^0. Remember to record the boiling point range of the fraction you collected, i.e. the temperature at which distillation of the fraction started and finished. Record the weight of the product both crude and pure product after distillation and store it in a clearly labeled sample bottle. (N.B. the sample bottle must be properly stoppered or the product which is very volatile, will evaporate! The sample should be stored in the refrigerator.

Questions

1. The chlorides of tertiary aliphatic alcohols are readily prepared by the action of conc. hydrochloric acid upon alcohol at room temperature, whereas primary aliphatic alcohols require sulfuric acid and the corresponding sodium halide. Explain the reason for this difference.
2. (a) Predict the order of reactivity of the following compounds towards reaction with silver nitrate in aqueous ethanol:

   $PhCH_2Cl$, $O_2N\text{-}C_6H_5\text{-}CH_2Cl$, $PhCl$

   $PhCH_2Br$, $CH_3CH_2CH_2CH_2Cl$

   (b) Predict the order of reactivity of the following compounds towards reaction with sodium iodide in anhydrous acetone:

$CH = CH\ CH_2Cl$, $CH_2 = CH\ CH_2CH_2Cl$, $CH_2 = CH\ Cl$, $CH_3CHClCH_3$
Give reasons for your predictions.

3. Indicate, with reasons, whether the following compounds would be expected to undergo predominant S_N1 or S_N2 type attack.
   (a) $C_6H_5\ CH_2\ Br$ (b) $C_6H_5\ CH(Br)\ C_6H_5$ (c) $C_6H_5\ CO\ CH_2\ Br$

Base – Catalyzed Condensations

Reference:

1. C. D. Gutsche, 'The Chemistry of Carbonyl Compounds' Morri

son and Boyd, Chapter 30; Roberts and Caserio, p. 479 et seq.

2. Crouch, R. D.; et al., The Aldol Addition and Condensation: The Effect of Conditions on Reaction Pathway. J. Chem. Educ. 84 (2007); 475-476.

4.3 Preparation of Chalcone (Claisen-Schmidt Reaction)

In the presence of a base, aldehydes condense with compounds containing activated (acidic) methylene or methyl groups to give β-hydroxycarbonyl compounds

$$R-CHO+R^1CH_2COR^2 \rightarrow R-CH(OH)-CH(R^1)COR^2$$

An example of this reaction is the 'aldol condensation' in which two molecules of an aliphatic aldehyde react in presence of base to give a β-hydroxyaldehyde.

$$2R\ CH_2CHO \rightarrow R\ CH_2CH(OH)-CH(R)CHO$$

These β -hydroxy carbonyl compounds undergo ready dehydration to give α, β - unsaturated aldehydes and ketones which constitutes the best method of preparation of these derivatives. This dehydration reaction in special cases may occur spontaneously under the conditions of the original base-catalyzed condensation. For example, aromatic aldehydes and methyl ketones react in presence of base to give 1,3-diphenylprop-2-en-1-ones ('chalcones'). This is known as a Claisen-Schmidt reaction.

$$Ar\text{-}CHO + Ar^1\ CO.CH_3 \rightarrow Ar\text{-}CH{=}CH\text{-}COAr^1$$

Experimental Procedure

A solution of potassium hydroxide (0.14 g., 0.0025 mole) in water (2 ml) is added with stirring to a solution of an acetophenone (0.025 mole) in ethanol (25 ml). To this mixture is added slowly with stirring a solution of an aromatic aldehyde (0.025 mole) in ethanol (50 ml) and the mixture allowed to stand until precipitation of the product is complete (perhaps overnight). The crude chalcone, is filtered with suction on a sintered glass funnel, wash with cold water. Recrystallise the chalcone from rectified spirit warmed to 50 °C and filter the product. The yield of pure benzylideneacetophenone (a pale yellow solid), m.p. 56-57 ^{0}C and calculate the yield of pure product. This substance should be handled with great care since it acts as a skin irritant.

Questions

1. What is the function of the base in this reaction?
2. How can you conduct the following transformations?

   a) R – Br → $RCH_2CO\ CH_3$

   b) R – Br → $RCH_2\ CO_2H$
3. Reaction of propionaldehyde with aqueous sodium hydroxide solution affords the aldol (I). Why is the alternative aldol (II) not formed?

   $CH_3\ CH_2\ CH(OH)\ CH(CH_3)\ CHO$ (I) $CH_3\ CH_2\ CH(OH)\ CH_2CH_2\ CHO$ (II)

4.4 Electrophilic Addition to Alkenes

Alkenes are much more susceptible to electrophilic attack than alkynes, which react slowly or not at all with many reagents that readily attack carbon-carbon double bonds. The alkene double bond is an electron-rich system which readily enters into addition reactions with electron-seeking (electrophilic) reagents such as bromine.

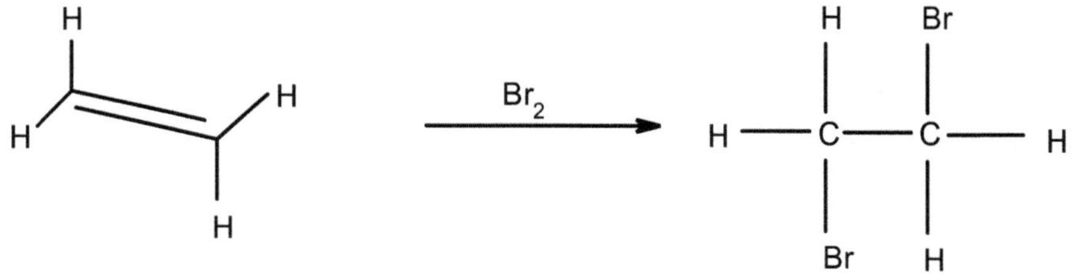

Bromination of alkenes is considered to involve interaction of the double bond with a polarized bromine molecule in a stepwise process. The commonest mechanism scheme is outlined below using ethylene.

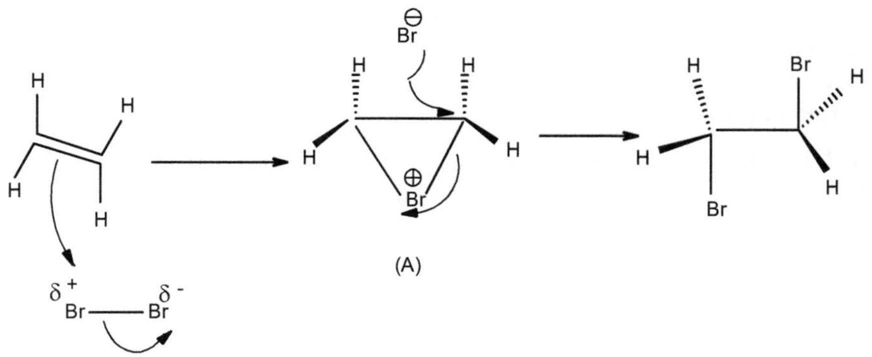

The first intermediate formed is thought to be a cyclic brominium ion (A). This intermediate has a three membered ring and the positive charge is on the bromine atom. Since bromine is a large atom this has the effect of blocking one side of the molecule; the only way the second bromine atom can attack to form a bond from the opposite side of the molecule. This is called *trans*-addition.

4.5 Bromination of a Chalcone (Benzylideneacetophenone)

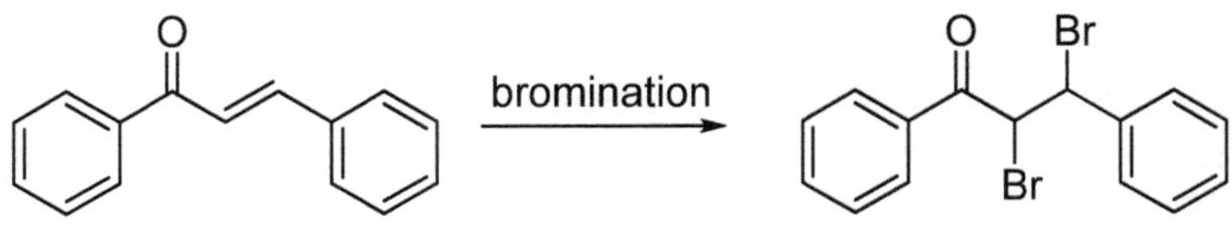

Dissolve pure chalcone, 1,3-diphenylprop-2-en-1-one (0.02 mole) in carbon tetrachloride (100 ml) and cool the solution in an ice-bath. To this solution, carefully add slowly with stirring a solution of bromine (Danger#) in carbon tetrachloride (0.02 mole). When addition is complete, the precipitated dibromide is filtered off and washed with two portions of hot carbon tetrachloride (small amounts each time). Crystallize the compound to constant melting point using a suitable non-hydroxylic solvent.

Check the purity of the product by TLC.

Dissolve a small amount of the final product in methylene chloride, and spot it on the plate. Also spot a dilute solution of the starting material, chalcone, on the same plate. Develop the plate in methylene chloride and use the UV lamp or iodine chamber to visualize the spots to see if there are any by-products or starting chalcone in your final product.

A solution of bromine in CCl_4 of known concentration will be provided by the demonstrator. Since this is extremely corrosive, great care should be exercised in its use. Rubber gloves should be used and the reaction should be carried out in a fume cupboard.

Q. 1 Compare the infrared spectrum of your product with that of the chalcone obtained in the previous experiment. What are the most important differences?

4.6 Preparation of Dibromomethyl Cinnamate

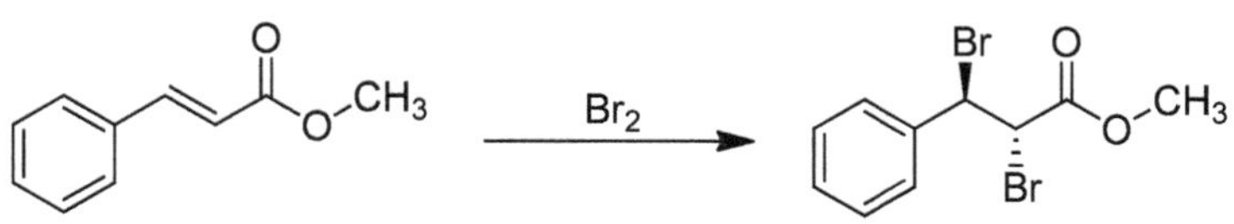

Experimental Procedure

Weigh 1 g of methylcinnamate and dissolve in dichloromethane (5 ml) in a small round-bottomed flask. Stopper lightly with a cork or stopper. In another small flask with a stopper dissolve 0.7 ml bromine (using pipette) in 5 ml of dichloromethane and keep it ready. Pipette 2 ml of this bromine solution and add to the cinnamate solution in one portion. (the preparation should be conducted in a fume cupboard because bromine should be handled with extreme care since it will cause severe burns and its vapors are irritating and poisonous). Re-stopper the flask, swirl gently to mix and leave to stand for 15 – 30 minutes. Distil off the dichloromethane on the steam bath. Don't forget to use boiling stones and a condenser. Crystallize the solid residue from petroleum ether (60-80 ^{0}C). Check the melting point, and calculate the percentage yield.

4.7 Relative Rates of Bromination of Aromatic Compounds

The rates of electrophilic substitution of aromatic compounds are greatly influenced by the groups present in the aromatic molecule. In this experiment, solutions containing equal concentrations of four different aromatic compounds in 90% acetic acid will be treated with equal amounts of bromine in the same solvent. The compounds to be tested are:

Benzene

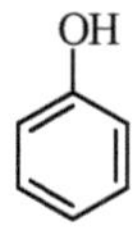

Phenol

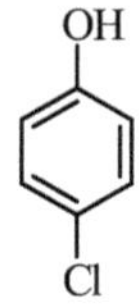

p-Chlorophenol

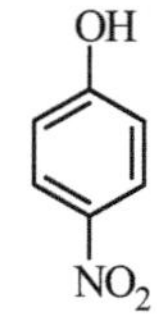

p-Nitrophenol

Experimental procedure

In four separate labeled test tubes place 2 ml of 0.2 M solutions (in 90% acetic acid) of each of the above compounds. Place the tubes in a 400 ml. beaker filled with warm water (25-27^0). Also place a test-tube containing 10 ml. of 0.05 M bromine solution (in 90% acetic acid) in the same water bath and allow the tubes to stand for 10 minutes until the contents have reached the bath temperature.

Add 2 ml. of bromine stock solution to each of the labeled test tubes, mix equally with a glass rod, and note the time (in seconds) for the solutions to decolorize. Record the times when the solutions become colorless or permanent faint yellow. If any of the solutions show no visible change after two hours, simply record the time as greater than 120 minutes.

Questions

1. From your results with benzene and phenol say whether the OH group is ring - activating or ring de-activating towards bromination in acetic acid. Why?
2. From your results with the three phenols say whether the Cl and NO_2 groups are ring activating or ring de-activating towards bromination in acetic acid. Explain the order observed with the three phenols.
3. Arrange the groups H, NO_2, Cl, OH in decreasing order of ring activation towards bromination in acetic acid
4. Draw the major monobromination product for each of the four compounds brominated.

4.8 Preparation of 2,4,6-Tribromoaniline

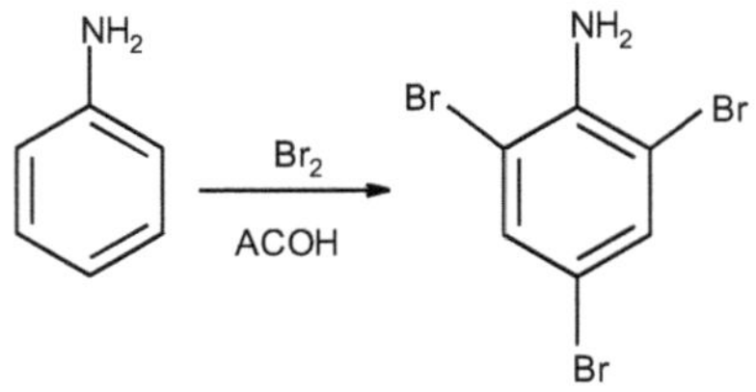

Chemicals: Aniline, bromine, acetic acid

Experimental procedure: Dissolve 1 g of aniline in 4 g of glacial acetic acid in a 200 ml beaker and while stirring add slowly a solution of 1.7 ml of bromine in 4 ml of glacial acetic acid. The beaker should be cooled in ice water during the addition as the reaction is exothermic, pour into excess of water, filter at the pump, wash well with water and dry. Recrystallise a small portion from rectified spirit. The yield of tribromoaniline is quantitative, m.p. 120°C.

4.9 Preparation of *p*-Bromoacetanilide

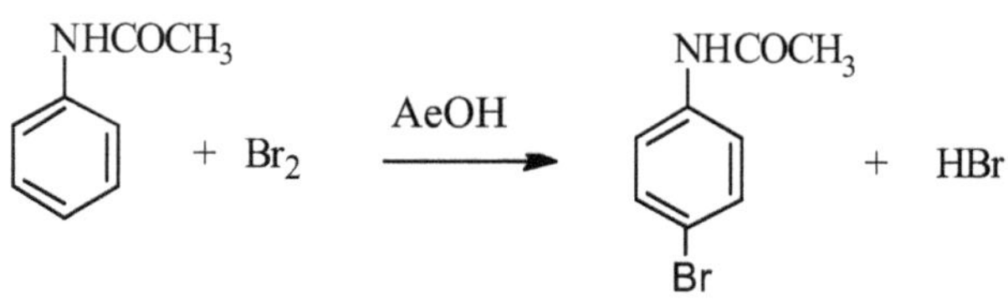

Chemicals: Acetanilide, bromine, glacial acetic acid

Experimental procedure: Dissolve 2 g of finely powdered acetanilide in 6 ml of glacial acetic acid in a 50 ml. conical flask. In another small flask dissolve 0.8 ml of bromine in 5 ml of glacial acetic acid, and add this solution slowly to the acetanilide solution, shaking the latter throughout the addition to ensure thorough mixing; stand the flask in cold water. Allow the final reaction mixture to stand at room temperature for 20 minutes. Pour the reaction mixture in a beaker containing water and if it is coloured, add just sufficient sodium bisulphite solution to remove the orange colour. Filter the crystalline precipitate with suction on a Buchner funnel, wash thoroughly with cold water, drain, and finally recrystallise from rectified spirit. The yield of p-bromoacetanilide is 2.5 g. m.p. 167°C.

4.10 Preparation of *p*-Bromoaniline (Hydrolysis)

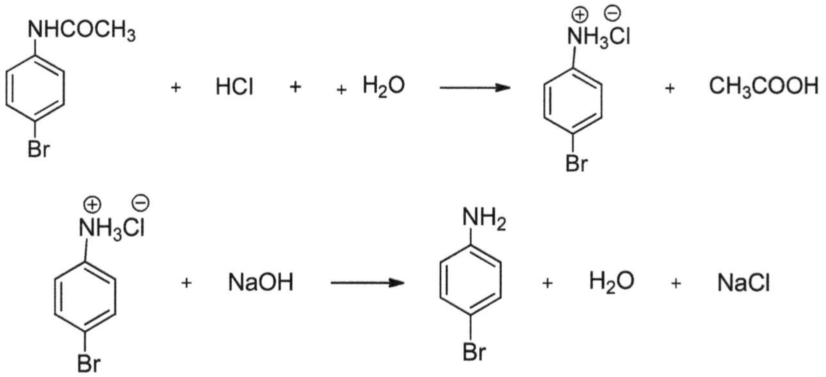

Chemicals: *p*-Bromoacetanilide, conc. hydrochloric acid.

Experimental procedure: In a 100 ml round-bottomed flask are placed 2 g of p-bromo acetanilide and 20 ml of 6N hydrochloric acid. The suspension is refluxed for about 15 minutes until a clear solution is obtained. When reaction is completed, the flask is cooled under a cold water tap. White crystals of *p*-bromoaniline hydrochloride will be deposited because of the limited solubility of the salt in hydrochloric acid solution. A cold solution of 6 g of sodium hydroxide in 10 ml of water is added in portions with cooling. The resulting alkaline solution is cooled. *p*-Bromoaniline separates as an oil, which soon crystallises. Filter the crystals at the pump, wash with cold water and dry in the air upon pads of filter paper. The yield is 1.6 g m.p. 66°C. Recrystallisation from dilute alcohol is usually unnecessary.

4.11 Aromatic Substitution: Nitration

4.11.1 Preparation of *ortho* and *para*-Nitro Phenol

Phenol undergoes electrophilic aromatic substitution with great ease. Thus, it can be nitrated by dilute nitric acid to give a mixture of *o*- and *p*-nitrophenols. These compounds can be readily separated because the former is steam volatile.

Mechanism

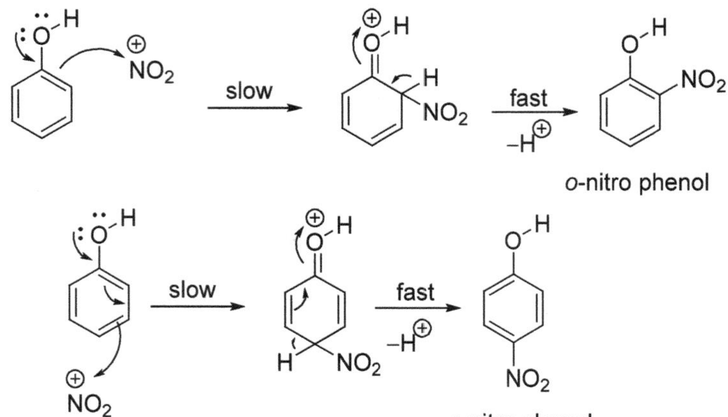

Chemicals: Phenol, conc. nitric acid.

Experimental procedure: Prepare a mixture of concentrated nitric acid (6.5 ml) and water (20 ml) in a 250 ml conical flask, warm a mixture of phenol (5 g) and water (4 ml) in another small conical flask until the phenol is molten. Shake vigorously to get an emulsion of phenol in water. Cool the nitric acid in ice-water and add the phenol so that the temperature does not rise above 30^0. When the addition of the phenol is complete, remove the flask from the cooling mixture and allow to stand for at least 2 hours. During the first hour of standing keep the thermometer in the liquid and do not allow the temperature to exceed 50-55 ^{O}C.

Add 50 ml of water and decant off the reddish–brown aqueous solution from the heavy black oil. Add 100 ml of water, shake, and again decant off the aqueous layer. Add 200 ml of water and steam distil the mixture. The *ortho*-nitrophenol is steam volatile and crystallises out in the cold distillate. If the *ortho*-nitrophenol crystallises in the condenser, run out the condenser water for a few minutes when the *ortho*-nitrophenol will melt and pass on into the distillate. When the distillation is complete, cool the distillate in ice-water and filter off the *ortho*-nitrophenol at the pump. Dry by pressing between sheets of filter paper.m.p. 46°C. The yield of *ortho*-nitrophenol 35%.

The p-nitrophenol can also be isolated from the residue in the distillation flask.

Isolation of *p*-nitrophenol

As soon as possible after the steam distillation, filter the hot residue through a conical funnel in which are two wet fluted papers. Discard the black residue; allow the filtrate to stand overnight. Collect the dark crystals, and suspend them in a 250 ml flask. to gether with cold water (50 ml), conc. hydrochloric acid (2.5 ml) and a small amount of decolourising charcoal. Fit a reflux condenser, bring to the boil (Bunsen flame), and boil for 10 minutes. Filter hot through a conical funnel, allow the filtrate to cool, and collect the *p*-nitrophenol as long brownish crystals, m.p. 112°C. yield 25%.

Mechanism

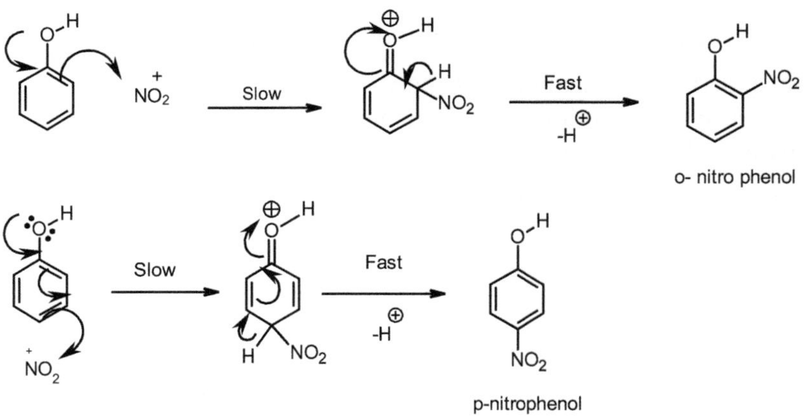

Questions

1. Why is o-nitrophenol steam volatile while *p*-nitrophenol is not?
2. Why is no *m*-nitrophenol obtained?

4.11.2 Preparation of *meta*-Dinitrobenzene

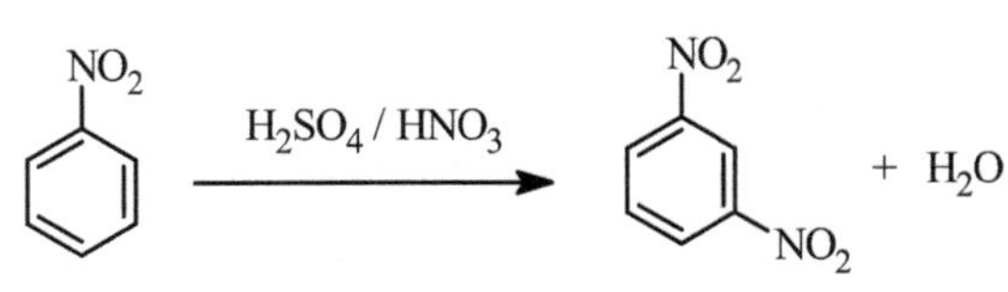

Chemicals: Fuming nitric acid, sulphuric acid, nitro benzene.

Experimental: This preparation of *m*-dinitrobenzene should be carried out in a fuming cupboard. Carefully mix together 5 ml of fuming nitric acid and 7ml of concentrated sulphuric acid in a 100 ml round-bottomed flask, and fit a reflux condenser to the flask. An apparatus with ground-glass joints should be used, fuming nitric acid rapidly attacks corks. Place a few pieces of pumice stone in the acid and then very slowly add 4.0 ml of nitrobenzene in 1 ml portions, pouring it down the condenser and shaking the flask well to ensure thorough mixing. Heat the mixture, with frequent shaking, on a boiling water bath for 30 minutes. After the flask has cooled, pour the contents slowly into about 150 ml of cold water in a beaker stirring thoroughly. The residual acid will dissolve in the water and m-dinitrobenzene will collect as an yellow solid. Filter using a Buchner funnel and filter pump, wash thoroughly with cold water and allow it to drain. Finally recrystallise the product from 30 ml of methylated spirit, to free it from traces of the *ortho* and *para*-isomers. The yield is 4.8 g m.p. 89-90°C.

4.11.3 Preparation of 1,4-Dichloro-2-Nitrobenzene

In this experiment *p*-dichloro benzene is reacted with fuming nitric acid in the presence of sulfuric acid. The presence of the two chlorine atoms deactivates the molecule to the extent that a reaction temperature of approximately 100 ^{o}C is required.

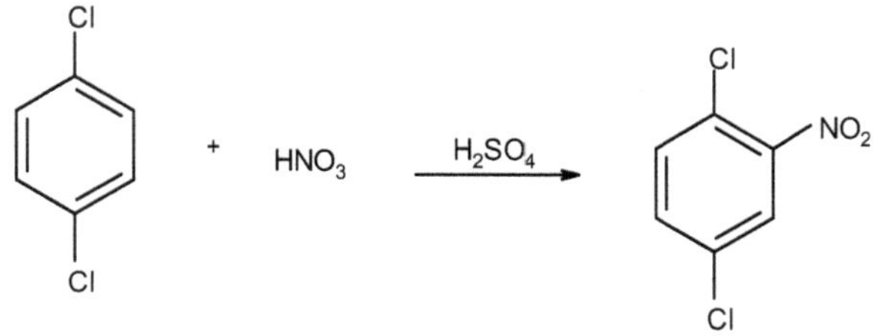

Experimental procedure: *p*-Dichlorobenzene (7.0 g) is placed in a 50 ml round-bottomed flask. A mixture of 3 ml fuming nitric acid (density 1.54 g/ml) and 3.5 ml of concentrated sulfuric acid is added in three portions. The flask is swirled vigorously between the additions. A condenser is attached to the flask in a reflux position, and the reaction mixture is heated for one hour over a boiling water bath. The reaction mixture is cooled to room temperature. Crystals of the product appear. The slurry is poured into 100 ml ice cold water. The suspension is filtered and the crystals of 1,4-dichloro-2-nitrobenzene are washed thoroughly with water on the filter. The product is recrystallised from ethanol. The recrystallisation solution is cooled slowly to prevent oiling of the low melting point solid. About 8.0 g of a light yellow solid, melting at about 54°C, should be obtained. (lit., m.p. 56°).

4.11.4 Preparation of *p*-Nitroacetanilide from Acetanilide

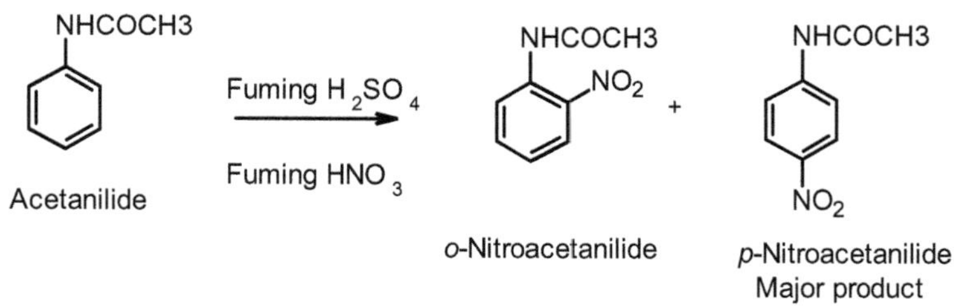

Chemicals: Acetanilide, conc. sulphuric acid, conc. nitric acid, acetic acid.

Experimental: Add 5 g of powdered acetanilide to 5 ml of glacial acetic acid contained in a 100 ml beaker and then to the well-stirred mixture add 10 ml of concentrated sulphuric acid. The mixture becomes warm and a clear solution results. The beaker is cooled with a freezing mixture of ice and salt until the temperature of the reaction mixture falls to 0-5°C. Now, whilst stirring the viscous mixture continuously with the thermometer, add 2 ml of concentrated nitric acid carefully, drop by drop from a dropper, while the temperature is maintained below 10°C. Remove the beaker from the freezing mixture, allow it to stand at room temperature for 30 to 45 minutes. Pour the contents on to 50 g of crushed ice (or 100 ml of cold water), whereby the crude nitroacetanilide is at once precipitated. Allow the mixture to stand for 15 minutes and then filter at the pump, wash thoroughly with cold water until free from acids and drain well. Recrystallise the crude pale yellow product from ethanol or methylated spirit, filter at the pump. wash with little cold spirit and dry in the air upon filter paper. The yellow o-nitro acetanilide remains in the filtrate whilst the *p*-nitro acetanilide is obtained as colourless crystals, m.p. 214°C. The yield is 4 g.

Questions

1. What is the function of sulfuric acid in this experiment?
2. Explain why *p*-xylene under goes nitration more readily than *p*-dichlorobenzene?

4.12 Prepartion of *m*-Nitroaniline from *meta*-Dinitrobenzene (Partial reduction)

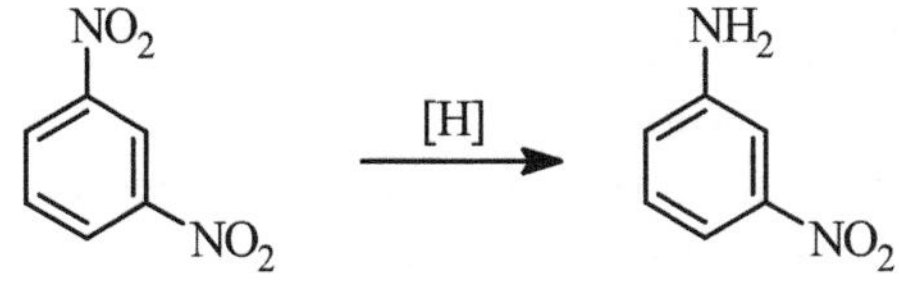

Chemicals: *m*-Dinitrobenzene, sulphur, sodium sulphide

Experimental procedure: Dissolve 5.4 g of crystalline sodium sulphide ($Na_2S.9H_2O$) in 15 ml of water in a 100 ml beaker and add 0.7 g of finely powdered sodium hydrogen carbonate with stirring. When the carbonate dissolved completely, add 14 ml of methanol and cool below 20°C

Filter off the precipitate sodium carbonate at the pump, using a small Buchner funnel. Wash the precipitate with 5 ml of methanol. Retain the filtrate and washings; these contain about 1.5 g of sodium sulphide in solution and must be used forth with for the reduction.

Dissolve 2.0 g of *m*-dinitrobenzene in 15 ml of hot methanol in a 100 ml round bottomed flask and add, with shaking, the previously prepared methanolic solution of sodium hydrogen sulphide. Attach a reflux condenser and boil the mixture for 20 minutes. Allow the reaction mixture to cool and fit the condenser for distillation. Distill off most of the methanol (30-35 ml) from a water bath. Pour the liquid residue with stirring into about 50 ml of cold water. Collect the yellow crystals of m-nitro aniline by filtration, wash with water and recrystallise from 75 percent aqueous methanol. The yield is 1.1 g, m.p. 114°C.

4.12.1 Preparation of *p*-Nitroaniline

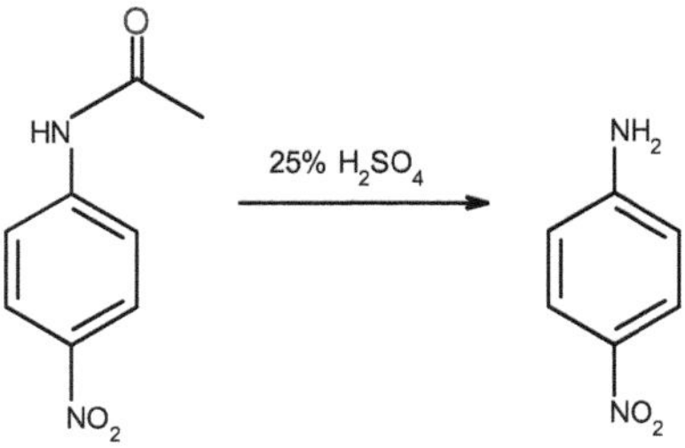

The hydrolysis of the acetanilide is best conducted using strong aqueous sulphuric acid prepared by adding very cautiously 20 ml of conc. H_2SO_4 to 15 ml of water or you can prepare 70% acid*

Place 3.5 g of p-nitroacetanilide in a 50 ml round-bottomed flask and add 20 ml of the aqueous sulphuric acid. Fit a reflux condenser, add a boiling stone to the solution, and reflux the mixture gently over a small Bunsen flame in the fume-cupboard for 20 minutes. Pour the hot, clear solution into 100 ml of cold water in a beaker and make the solution alkaline by the addition of dilute sodium hydroxide solution. Yellow crystals of p-nitro aniline should separate. Cool the mixture thoroughly to room temperature and collect the crystalline solid using a Buchner flask and funnel. Wash the product thoroughly with cold water and dry. The yield is 1.5 g.

Recrystallise the material from minimum volume of a hot mixture of equal parts of methylated spirit and water. The hot clear solution should be boiled on the steam bath for a few minutes with a small quantity of charcoal. The charcoal is then removed by filtering the hot solution through a preheated glass funnel and fluted filter paper. Cool the filtrate when the pure p-nitroaniline separates as yellow crystals. Collect these in the normal way, dry in a desiccator. It has m.p.148^{0}C.

***Note:** A 70% sulphuric acid solution is prepared by adding 60 ml of conc. sulphuric acid cautiously in a thin stream with stirring to 45 ml of water with cooling.

4.13 Friedel-Crafts Alkylation: Preparation of *p-tert-B*utylphenol

Aromatic compounds may be alkylated or acetylated in the Friedel-Crafts reaction by alkyl and acyl halides respectively in the presence of a Lewis acid such as aluminium chloride. An example of this is the preparation of *p-tert-*butylphenol from phenol and t-butyl chloride.

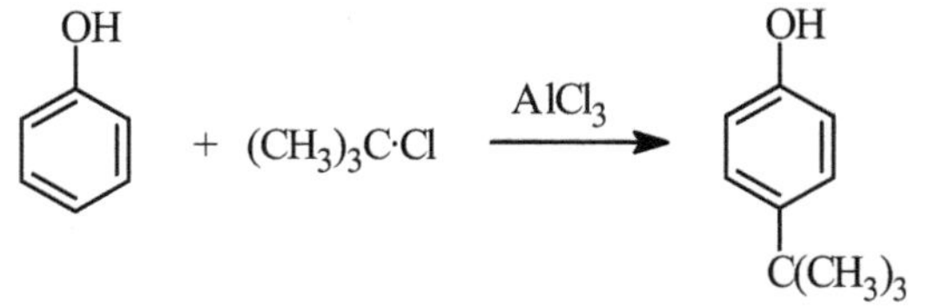

The strong *ortho/para* directing effect of the hydroxyl group and the large size of the *tert*-butyl group results in the almost exclusive formation of the *para* isomer.

Experimental procedure

Place *tert*-butyl chloride (5.5 g., 6.5 ml) and phenol (4.7 g) in a 50 ml conical flask and stir until the phenol is nearly dissolved. Place the flask in a fume cupboard and add aluminium chloride (0.5 g) a few pieces at a time. Hydrogen chloride evolution should start immediately. If the reaction mixture becomes warm, cool the flask for a few minutes in a beaker of cold water. Use the glass rod to stir the contents of the flask occasionally to expose new surfaces of the catalyst. After 20-30 minutes, the contents of the flask should solidify.

When the reaction mixture is essentially solid, add 25 ml. of water and break up the solid gently with a spatula until it is finely divided. Filter off the solid. Press it as dry as possible with a spatula or a clean cork and allow air to be drawn through it for about five minutes.

Transfer the solid to 250 ml. conical flask and add enough petroleum ether b.p. 60-80^0 to dissolve the solid (i.e. minimum quantity of boiling solvent), using charcoal (about 1 spatula-load). Filter the hot solution rapidly through a preheated funnel. *p-tert*-butyl phenol will crystallize from the filtrate. It has m.p. 99 ^{0}C.

Question

1. Outline one other method for the preparation of *p-tert*-butyl phenol from *t*-butyl benzene.

4.14 Oxidation

Since a diverse range of oxidizing agents are available to organic chemists, the value of a particular reagent lies in its selectivity. In other words, the reagent of choice for a particular reaction is one which will oxidize one functional group in the molecule while other potentially oxidizable groups present are not affected. One of the mildest oxidizing agents is manganese dioxide which is prepared in active form by the thermal dehydration of manganous hydroxide. This reagent will selectively oxidize compounds containing activated methylene or hydroxy methylene groups to ketones.

4.14.1 Oxidation of 4,4'–Dimethylbenzhydrol to 4,4'-Dimethylbenzophenone

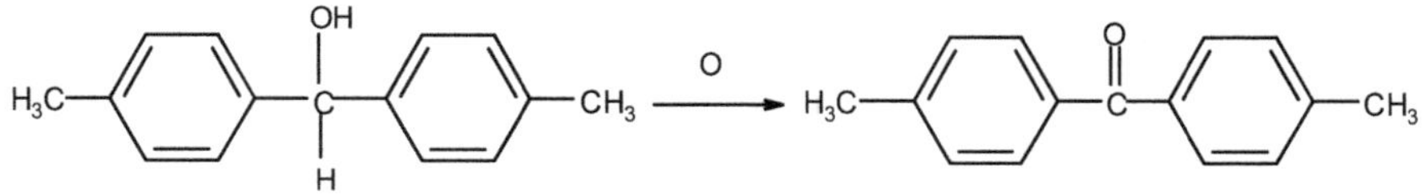

Experimental procedure: Dissolve 4,4'- dimethyl benzhydrol (1 gm.) in cyclohexane (30 ml) and place the solution in a 100 ml round-bottom flask fitted with reflux condenser and calcium chloride drying tube. To the solution, add manganese dioxide (4 gm.) and heat the mixture on a steam bath until the solvent refluxes gently. (N.B the apparatus must be securely clamped since the mixture 'bump' while boiling). At intervals of 30 minutes remove an aliquot (ca. 1 ml) of the solution, filter into a test tube and examine the filtrate by thin layer chromatography using a solution of 4,4'-dimethylbenzhydrol in toluene for comparison. Use toluene as the eluant. Observe the production of 4,4'-dimethylbenzophenone in the reaction by developing the chromatograms in iodine vapour. When TLC indicates that the oxidation is essentially complete, allow the reaction mixture to cool and filter off the manganese dioxide. Evaporate the filtrate to dryness and crystallize the residue from methanol, m.p. 93 ^{0}C.

4.14.2 Preparation of Organic Acid - Benzoic Acid from Toluene

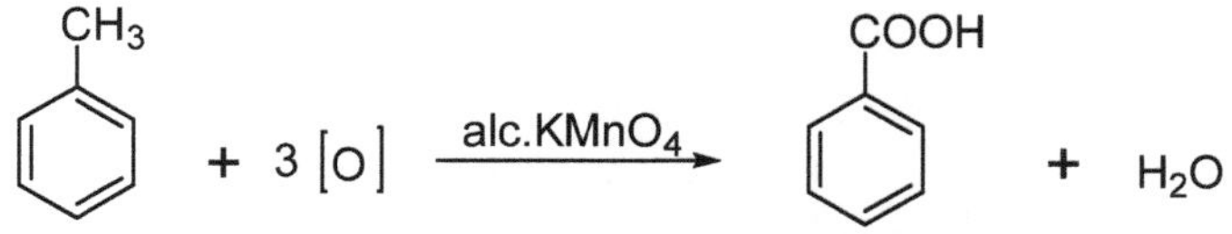

Benzoic acid

Experimental procedure: The aromatic ring is stable towards most oxidizing agents, and it is therefore possible to oxidize a side-chain to a carboxylic acid group without affecting the aromatic nucleus.

Method 1: Place potassium permanganate (4 g) and sodium carbonate (o.5 g) in a small round-bottomed flask. Add water (40 ml) and warm gently to dissolve the solids. Cool the solution, add toluene (1.25 ml) and a boiling stone. Boil the mixture gently under reflux over a low Bunsen flame or oil bath for 60-90 minutes. Swirl the flask occasionally to minimize bumping.

Cool the mixture to room temperature, acidify with dilute sulphuric acid (30-35 ml) and check that the solution is acidic using indicator paper. Add, in small portions, sufficient solid sodium matabisulphite (3.5 g) to destroy the excess of permanganate and manganese dioxide. At this point the colorless solution should contain, in suspension, white crystals of benzoic acid.

Cool the mixture thoroughly, collect the crystals by filtration, and recrystallise the product from the minimum of boiling water. Filter the product, dry and weigh the compound, m.p. 121 ^{0}C. Using the above equation calculate the percentage yield of benzoic acid. (Specific gravity of toluene = 0.87).

Method 2: Place about 5 ml of toluene and 30-40 ml of water in a small round-bottomed flask and add about 1 g of solid potassium permanganate and 5 ml sodium hydroxide solution. Place a reflux condenser in the neck of the flask, and boil the contents for 3-4 hours or until no purple color remains, but only the dark brown colour of manganese dioxide. Filter off the solid manganese dioxide, separate the lower aqueous layer from the unchanged toluene using a tap funnel, and acidify the aqueous layer with dilute hydrochloric acid (10 ml should be sufficient). The white crystals which separate are of benzoic acid. These should be filtered, washed with a little cold water, dried by pressing between filter-papers. m.p. 121^{0}C.

Ref: Organic chemistry by B. J. Stokes, Edward Arnold (Publishers Limited), 1972.

4.14.3 Preparation of *p*-benzoquinone

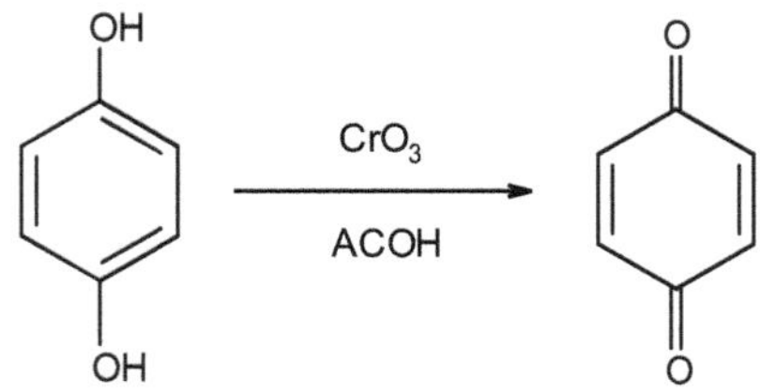

Experimental procedure: Cool a solution of 3.3 g of hydroquinone in 15 ml of 60 percent acetic acid containing in a 150 ml beaker to below 5^0 C in an ice bath. Dissolve 4.2 g of chromium trioxide in 10 ml of water, and add 3 ml of glacial acetic acid drop by drop at such a rate that the temperature does not rise above 10 ^{0}C. Filter the mixture at once and wash the quinone two or three times with small amount of ice-cold water. Spread the material upon filter paper until dry, but no longer or the quinone will be lost through sublimation. The yield of quinone (a bright yellow crystalline solid), m.p. 115 ^{0}C, is 2.1 g (66%), it darkens when exposed to light.

Impure quinone may be purified by placing it in a distilling flask attached to a condenser and passing a rapid current of steam into the flask: the p-benzo Quinone sublimes and collects in the receiver. It is separated from the water by filtration and dried; the m.p. is 116 ^{0}C. The vapour has a penetrating odour and attacks the eyes.

4.15 Reduction

Selective reduction with sodium borohydride

Object: To reduce a carbonyl group selectively in the presence of a nitro group; the conversion of 4-nitrobenzaldehyde into 4-nitrobenzyl alcohol.

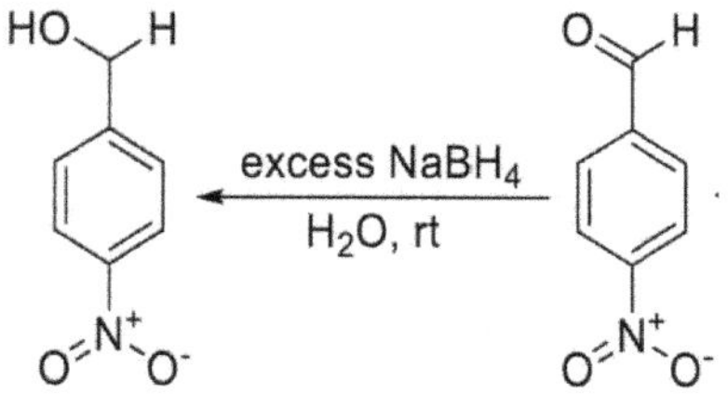

Background: There are many non-catalytic reducing agents which show some specificity for particular compounds containing the carbonyl (C=O) group. These reducing agents are mostly based upon metal hydrides and effectively act as sources of hydride ion (H^-).

Lithium aluminium hydride ($LiAlH_4$), more properly called *lithium tetrahydroaluminate*, is the most powerful. It will reduce all types of carbonyl groups, e.g. in esters, acids, amides, ketones, etc. The reactivity of lithium aluminium hydride can be modified to advantage in several ways. For example, it will react readily with three equivalents of t-butyl alcohol to give a sterically hindered hydride of much reduced reactivity:

$$LiAlH_4 + 3Bu^tOH \rightarrow LiAlH(OBu^t)_3 + 3H_2$$

This hydride will react with acid chlorides to give aldehydes with no further reduction. Lithium aluminium hydride on the other hand, converts acid chlorides to primary alcohols:

$$R-COCl + LiAlH(OBu^t)_3 \rightarrow R-CHO$$

$$R-COCl + LiAlH_4 \rightarrow R-CH_2OH$$

Sodium aluminium hydride ($NaAlH_4$) reacts like $LiAlH_4$ with carbonyl compounds, but when reacted with monomethyl ether of ethylene glycol it yields a hydride ($NaAlH_2(OCH_2CH_2OCH_3)_2$ of reduced reactivity. Unlike lithium hydride this dihydride will reduce the carbonyl group of α, β - unsaturated ketones without affecting the olefinic bond.

$$R-CH=CH-CHO + LiAlH_4 \rightarrow R-CH_2-CH_2-CH_2OH$$

$$R-CH=CH-CHO + NaAlH_2(OCH_2CH_2OCH_3)_2 \rightarrow R-CH=CH-CH_2OH$$

Lithium aluminium hydride on its own will usually reduce aldehydes, acids and esters to primary alcohols, and to ketones to secondary alcohols. When an equimolar amount of anhydrous $AlCl_3$ is added to the reaction, the corresponding hydrocarbons rather than alcohols are produced. For most aldehydes and ketones the yields of hydrocarbon are superior to those obtained by Clemmensen reduction (HCl + Zn/Hg):

$$R-CHO + LiAlH_4 / AlCl_3 \rightarrow R-CH_3$$

$$R-CHO + LiAlH_4 \rightarrow R-CH_2OH$$

Lithium aluminium hydride will also reduce the N=O bond of nitro compounds, ultimately to give amines.

$$R-NO_2 + LiAlH_4 \rightarrow R-NH_2$$

Various hydro aluminates act as hazardous materials due to their high reactivity towards oxygen and moisture.

Because of their relatively high reactivity towards moisture and oxygen, the various hydroaluminates are hazardous materials.

Sodium borohydride (sodium tetrahydroborate) is not so reactive because boron is less electropositive than aluminium, and therefore the B-H bonds are less polarized than Al-H bonds. Sodium borohydride ($NaBH_4$) is a selective reducing agent for various carbonyl groups. Thus, it will not react with esters or carboxylic acids, but reduces ketones to secondary alcohols and aldehydes to primary alcohols. For these transformations sodium borohydride is to be preferred over lithium aluminium hydride ($LiAlH_4$).

$$R-CHO+NaBH_4 \rightarrow R-CH_2OH$$

Sodium borohydride does not attack olefinic double bonds even when conjugated with carbonyl groups, nor does it attack nitro groups. In this experiment you will use $NaBH_4$ to carry out a selective reduction of an aldehyde group in the presence of a nitro group.

Mechanism: The reduction process occurs in two stages: nucleophilic addition of the H-B bond across the polarized carbonyl group, followed by hydrolysis. The first stage can continue until all B-H bonds are replaced by B-O bonds:

$$BH_4^- + R-\overset{\overset{\displaystyle H}{|}}{C}=O \rightarrow R-CH_2-O-BH_3^- + 3RCHO \rightarrow (R-CH_2-O)_4B^-$$

The resultant borate ester anion is then hydrolyzed by aqueous acid.

$$(R-CH_2-O)_4B^- + H_3O^+ \rightarrow 4R-CH_2-OH$$

Note that one mole of $NaBH_4$ is reducing four moles of aldehyde.

4.15.1 Preparation of 4-Nitrobenzyl Alcohol

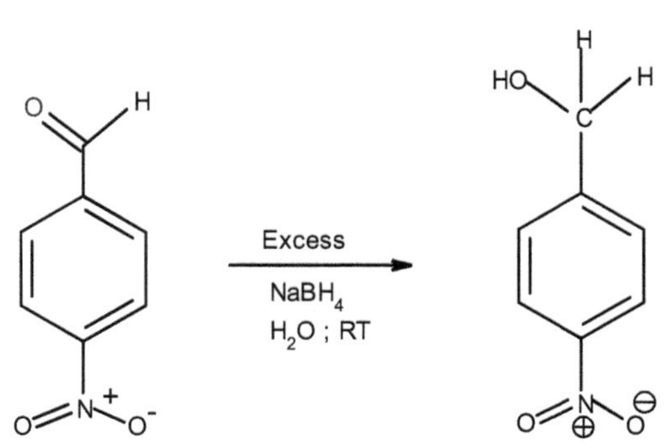

Experimental procedure

Dissolve *p*-nitrobenzaldehyde (0.50 g., 0.003 mole) in dry methanol (30 ml) contained in a 100 ml round-bottom flask fitted with calcium chloride drying tube. To this solution add sodium borohydrade (0.28 g., 0.008 mole) in two portions, and shake gently to encourage thorough mixing of the reactants. Gas will be liberated initially, but after approximately 15 minutes the reduction should be complete. Add water (100 ml) containing dil. HCl (10 ml). Wash the combined extract with water dry over anhydrous sodium sulphate, filter off the desiccant on a Buchner funnel. Examine a small portion of the filtrate by of the TLC to determine the extent of the reaction. Use 5% ethyl acetate in toluene as an eluant. Evaporate the filtrate to dryness. Recrystallise the residue from a mixture of light petroleum (b.p. 60-80^0) and methylene chloride and determine the yield and the melting point of product (90-93 ^{0}C), Yield 0.30 g.

Percentage yield = (Wt. product/ Wt. starting material X M.W. starting material/M.W product) $\times$ 100

Questions

1. How the following selective reductions can be carried out in the laboratory? Give adequate reasons for your choice of reagents.

   a) $CH_3CH = CHCOCH_3 \rightarrow CH_3CH_2CH_2COCH_3$

   b) $CH_3COCH_2CO_2C_2H_5 \rightarrow CH_3CH(OH)CH_2CO_2C_2H_5$

   c) $PhCO_2Et \rightarrow PhCH_2OH$

4.16 Acetylation

Acylation of an aromatic primary or secondary amine may be readily achieved by using acid chloride in the presence a base; however, acetylation is more usually carried out with acetic anhydride rather than the more offensive acetyl chloride.

4.16.1 Preparation of D glucose Pentaacetate from Glucose (acetylation)

The acetylation is in almost all cases affected with acetic anhydride using either the alkaline or an acidic catalyst.

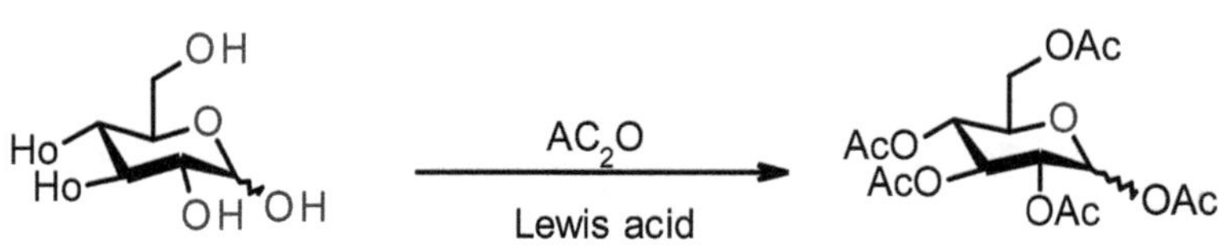

D-glucose pentaacetate

Experimental procedure: Place 1 g of anhydrous D-glucose in a 50 ml round bottomed flask, add 0.8 g of anhydrous sodium acetate and 6.0 ml of acetic anhydride. A reflux condenser is attached, and the flask is continuously heated with a flame to initiate the reaction as the unit is swirled to distribute the solid particles of the sugar, continue heating for a further 2 hours till a clear solution is obtained and then pour the reaction mixture into 100 ml of ice water with vigorous stirring. The product separates as an oil, but it crystallizes readily upon rubbing the gummy solid against the sides of the beaker with a stirring rod. Filter off the crystals, wash well with cold water and recrystallise from industrial spirit or methanol. The melting point is 132°C. The yield is 1.25g.

(**Note:** An excess of acetic anhydride is hydrolyzed by cold water, from which the product crystallizes.)

4.16.2 Preparation of Acetanilide

Acetanilide is conveniently prepared by the action of a mixture of acetic acid anhydride and glacial acetic acid on aniline.

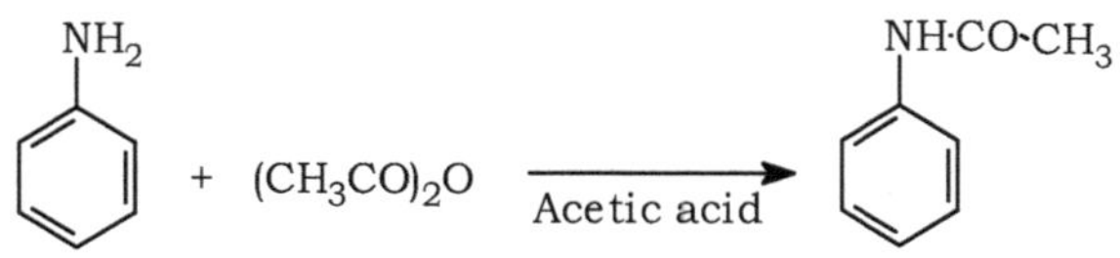

Experimental procedure: Carefully add 5 ml of aniline to a mixture of 5 ml glacial acetic acid and 5 ml of acetic anhydride contained in a dry round – bottomed flask equipped with a reflux condenser. Mix the contents and drop in two or three pieces of pumice stones to prevent bumping, and boil the contents of the flask for 30 minutes. At the end of this time pour the contents of the flask into 100 ml of cold water, stirring the water vigorously. As soon as the acetanilide crystallizes, filter using a Buchner flask and wash the crystals with water, and recrystallise from boiling water (30 ml) or aqueous ethanol, m.p.113 ^{o}C. The yield of acetanilide is 5g.

Mechanism

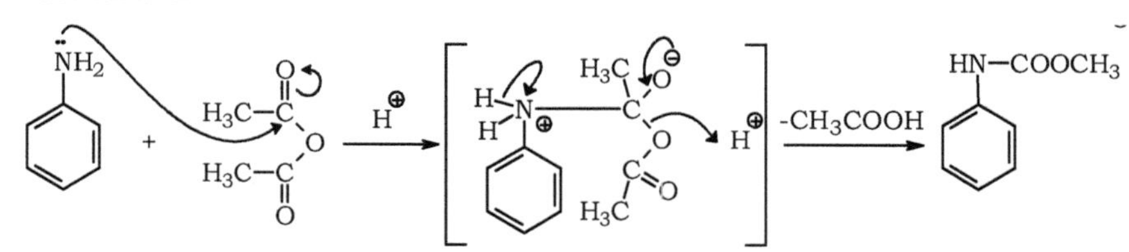

4.16.3 Preparation of "Aspirin" – Acetylsalicylic Acid

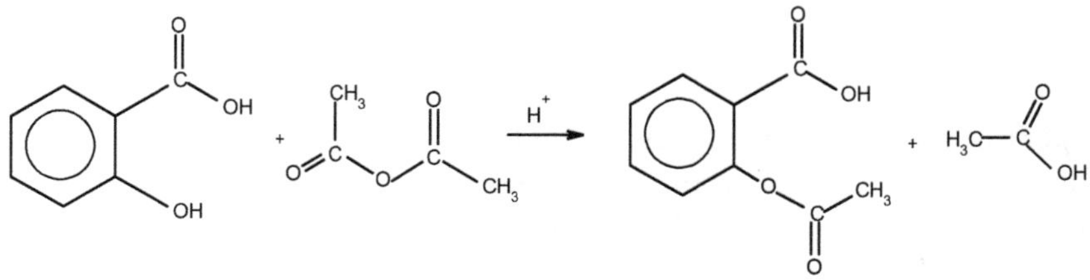

Experimental procedure: Place powdered salicylic acid (1 g.) in a 50 ml. conical flask and add acetic anhydride (2 ml.) using the liquid to wash down any material adhering to the flask walls. Note that there is no visible reaction. Cautiously add two drops of conc. sulfuric acid (H_2SO_4) and rotate the flask for a few minutes to mix the contents and dissolve the salicylic acid. Allow the mixture to stand for 5-10 minutes and, whether or not crystals of the acetate have begun to separate, rub the mixture well with a stirring rod and cool in ice-water until a stiff crystalline paste is obtained. Add 20 ml of water, rub to a thin paste, cool it in an ice-bath and collect the white powder by filtration through a Buchner flask.

Dry the material in a desiccator and record the weight and melting point of the crystals. Acetylsalicylic acid tends to decompose when heated, and it does not therefore have a clearly defined melting point. Decomposition points ranging from 128-137^0 are recorded in the literature.

Q. Calculate the percentage yield of acetyl salicylic acid based on the quantity of salicylic acid used.

4.17 7-Hydroxy-3-methyl Flavone

Ref. Indian J. Chem., page no. 696, 1968.

Aim: To synthesize Flavone by modified Baker-Venkataraman transformation:

It consists of two steps:

(i) Preparation respropiophenone from resorcinol

(ii) Preparation of 7-hydroxy-3-mythyl flavone

Step(i):Preparationof2,4-dihydroxy-propiophenone(Respropiophenone)

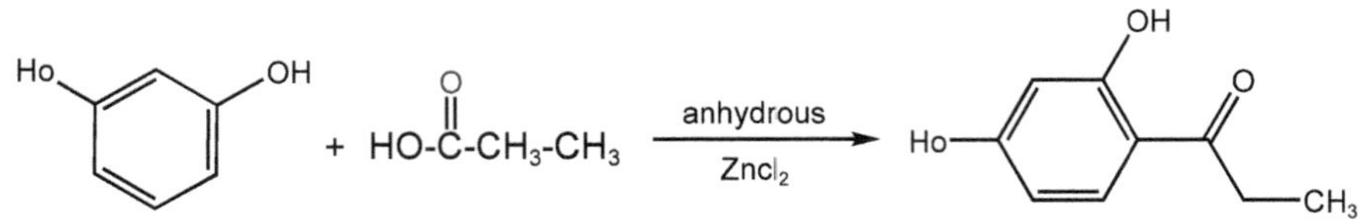

Experimental procedure: Resorcinol (5.5 g) was added slowly to a mixture of anhydrous $ZnCl_2$ and propionic acid taken in a 250 ml beaker while heating on a sand bath at 80 ^{0}C. After the addition is completed, heat the mixture slowly to 110 ^{0}C and rapidly to 140 ^{0}C for 45 minutes. Cool the red viscous mass obtained to room temperature and stirred in ice cold water. The orange red solid of respropiophenone is filtered and dried. Recrystallise from boiling water. The yield is 4 g. m.p. 101 ^{0}C.

Step (ii): Preparation of 7-hydroxy-3-methyl flavone

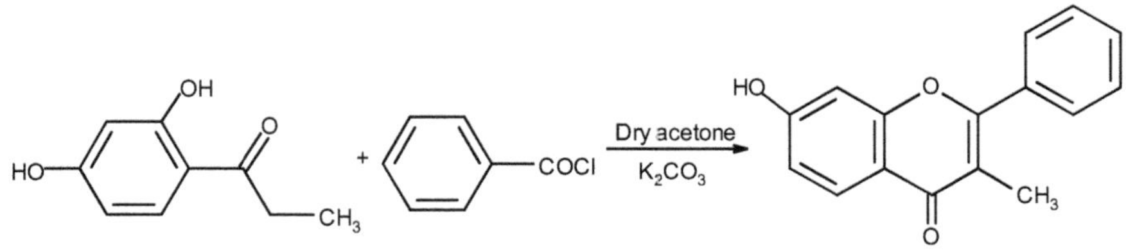

Experimental procedure: Place 4.0 g of respropiophenone, 100 ml of dry acetone, 2.1 ml benzoyl chloride and 10 g of anhydrous potassium carbonate in a 100 ml round- bottomed flask fitted with a reflux condenser and the solution is refluxed for 12 hrs on a water bath. The solvent was removed, the residue was cooled and treated with ice-cold water (100 ml). The reddish brown solution was saturated with CO_2 and product was separated by filtration. Further recrystallisation from methanol gives pure product (1.4 g) m.p.208 ^{0}C.

4.18 Preparation of 7-Hydroxy Coumarin (Pechmann Synthesis)

Ref: Experimental organic chemistry by David Todd. P-201

Principle: Condensation reaction of resorcinol with malic acid in the presence of sulphuric acid involves the loss of one carbon atom as carbon monoxide to give 7-hydroxy coumarin.

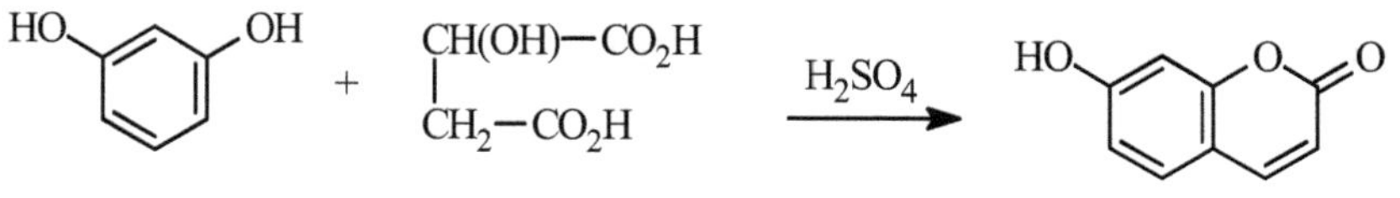

Experimental procedure: In a 250 ml conical flask place 2 g of resorcinol,

2.5 g of malic acid and 5 ml of concentrated sulphuric acid (carefully). The flask is held with a clamp, and is gently swirled by hand over the yellow flame of a Bunsen burner in hood. The heating is continued until the mixture begins to foam. By cautious heating maintain the foaming for several minutes. Let it cool in the air for about 5 minutes and with good swirling pour into about 100 ml of water containing some ice. Use little more water for rinsing. The cold suspension is suction filtered. Recrystallise the crude product from ethanol using a small amount of charcoal and the hot filtered ethanolic solution is ice cooled. Filter off the 7-hydroxy coumarin, the yield after air drying is 2 g., m.p. 227-228^{0}C.

4.19 7-Hydroxy-4-methyl Coumarin (Pechmann Synthesis)

It is a simple synthesis of coumarins involving the condensation of phenols with malic acid or β-ketoesters in the presence of conc. sulphuric acid or other condensing agents.

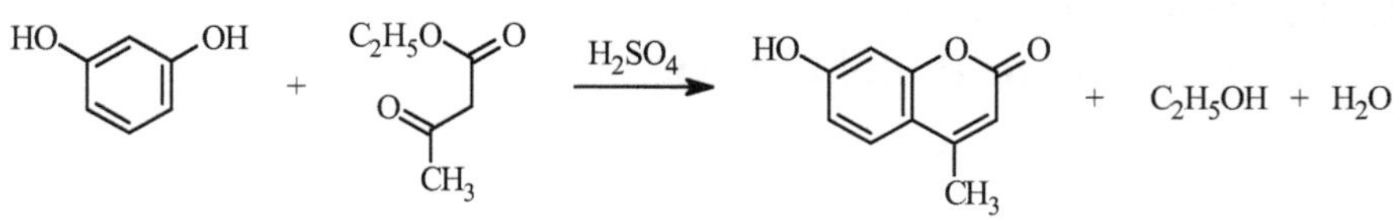

Chemicals: Resorcinol 1 g; ethyl acetoacetate 1.3 ml conc. sulphuric acid 10 ml.

Place 10 ml of concentrated sulphuric acid in a 150 ml conical flask with external ice-water cooling until the temperature of the acid is below 10°C. When the temperature falls below 10°C, add a solution of 1 g of resorcinol in 1.3 ml of ethyl acetoacetate, prepared in another test tube, drop wise, so that the temperature of the mixture does not rise above 10°C. Keep the reaction mixture at room temperature with occasional stirring. Pour the contents into 100 ml ice-water contained in a beaker. Collect the precipitate by suction filtration and wash it with cold water. For purification, dissolve the solid in 20 ml of 5% aqueous sodium hydroxide solution, filter and reprecipitate it by addition of dilute hydrochloric acid with stirring until the solution is acid to litmus. Collect the crude product by filtration at the pump, wash with cold water and dry. Recrystallise from ethanol or methylated spirit, using charcoal if necessary. The yield of 7-hydroxy-4-methyl coumarin is 1.5 g, m.p. 185 ^{O}C.

4.20 The Schotten-Baumann Reaction (Benzoylation)

4.20.1 Preparation of Benzanilide from Aniline

The benzoyl derivative of aniline can be prepared using benzoyl chloride according to the equation.

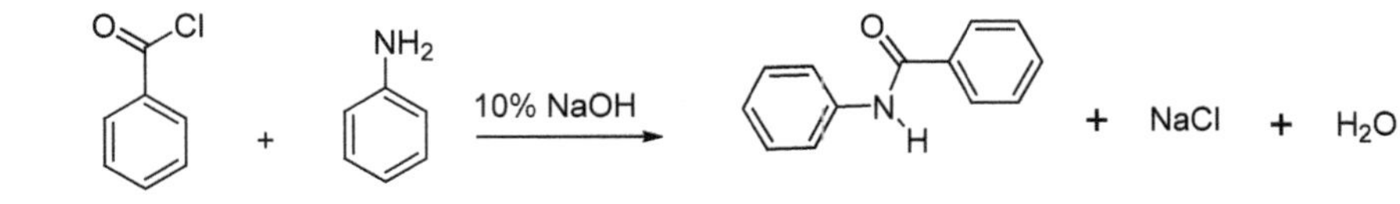

Benzoyl Chloride Aniline Benzinilide Sod.Chloride Water

Chemicals: Aniline, benzoyl chloride, sodium hydroxide.

Experimental: Add 2.7 ml of benzoyl chloride to 1.5 g of aniline in a small conical flask. Add sufficient dilute sodium hydroxide solution to make the mixture alkaline warm gently for 5 to 10 minutes on the steam bath. Test occasionally to make sure that the mixture is still alkaline if necessary add more alkali, cool, pour into water, filter the crystals and wash with water. Recrystallise the benzanilide from aqueous ethyl alcohol. Filter the crystals and dry them. Weigh the dry crystals, determine the melting point, m.p. 163°C. Calculate the percentage yield using the equation given above.

4.20.2 Preparation of Sulphanilic Acid from Aniline (Sulphonation)

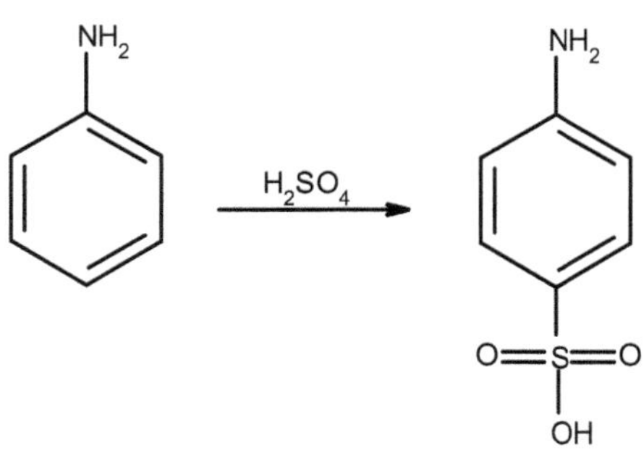

Experimental: Place 5.1 g (5 ml) of aniline in a 100 ml round-bottomed flask and slowly add 10 ml of concentrated sulphuric acid in small portions, shake the mixture gently during the addition and keep it cool by occasionally immersing the flask in cold water, and heat the contents at 180-190°C (fume cupboard) for about 5 hours. Allow the product to cool and pour it carefully with stirring into 100 ml of cold water or of crushed ice. Allow to stand for about 5 minutes and collect the crystalline solid on a Buchner funnel, wash it with water and drain. Purify the crude sulphanilic acid by recrystallising from about 100-125 ml of boiling water; if the resulting solution is coloured, add about 1 g of animal charcoal and boil for 10 minutes. Filter through a hot-water funnel or through preheated Buchner funnel. Dry the crystals between sheets of filter paper. The yield of sulphanilic acid is 5 g. The acid does not melt sharply and it decomposes on being heated. m.p.288 ^{0}C

4.21 Methyl 2-Naphthyl Ether (Methylation)

Chemicals required: 2-Naphthol (2 g), dimethyl sulphate (1.4 ml), 10% sodium hydroxide solution (6 ml).

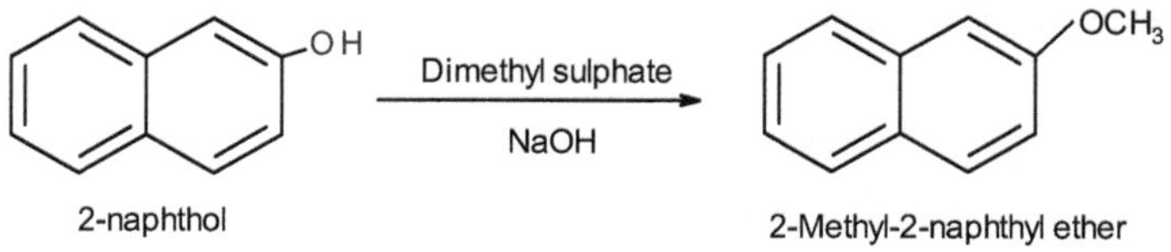

Experimental: In a 100 ml flask place 2-naphthol (2 g) 10% sodium hydroxide solution (6 ml) and shake the mixture. To this add dimethyl sulphate (1.4 ml) dropwise and warm the flask in a water bath at 70-80° for 1 hour to complete the reaction. Allow it to cool and the methyl naphthyl ether rapidly separates as a greyish-white powder. Filter the methyl ether at the pump, wash with water (20 ml) and drain.

Recrystallise from methylated spirit, from which the methyl 2-naphthyl ether separates readily as colourless crystals, m.p. 72°C. Yield is 1.8 g.

Note: Vapours of dimethyl sulphate highly poisonous. If the sulphate is splashed on the hands, wash immediately with plenty of ammonia solution followed by water.

4.22 Esterification of an Organic Acid

Carboxylic acids reacts with alcohols in the presence of a mineral acid catalyst to form esters. A general equation for the reaction is:

$$R-COOH+R'OH+H_2SO_4 \rightarrow R-COOR'+H_2O$$

4.22.1 Preparation of Methyl Benzoate

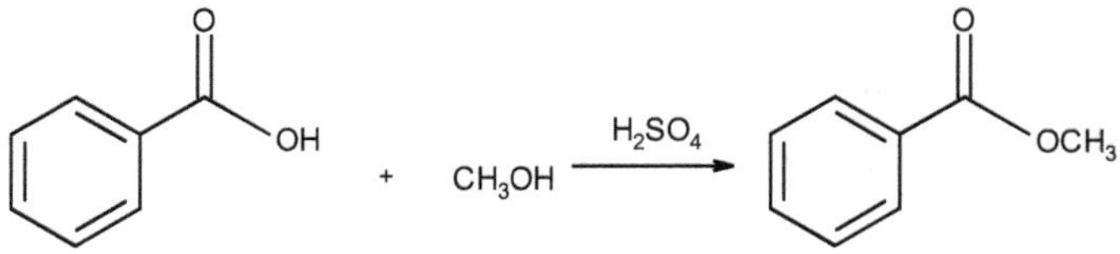

Place 2 g of benzoic acid in a small round-bottomed flask, and add 20 ml of methanol and 20 drops of concentrated sulfuric acid. Fit a reflux condenser and gently boiling the mixture on a steam bath for 45 min. Swirl the flask to ensure all the carboxylic acid dissolves. Cool the resulting solution, add 30 ml of water and place the resulting suspension in a separating funnel.

Add 50 ml of dichloromethane, stopper the funnel and shake it thoroughly.

Allow the two layers to separate, remove the stopper and run the lower dichloromethane layer in to a beaker. This extract contains the ester together with small amounts of unchanged organic acid. Re-extract the aqueous layer in the funnel with a further 25 ml dichloromethane. Separate the layers as before, combine the dichloromethane extracts, and discard the aqueous layer in the funnel.

Pour the combined dichloromethane into a clean separating funnel, add 50 ml of sodium bicarbonate solution, stopper the funnel and shake the mixture well.

N.B. Carbon dioxide may be evolved during the process and pressure in the funnel should be released from time to time by inverting the funnel and opening the stop-cock. Again, run off the lower dichloromethane layer (dichloromethane density is higher than water) into a beaker and discard the upper layer remaining in the funnel.

Pour the dichloromethane into a clean separating funnel, add 25 ml of water and shake. Separate the layers as before, running the lower dichloromethane into a small dry clean conical flask and dry the solution by addition of a little anhydrous sodium sulfate. Swirl and allow to-stand for 10 min. Filter the clear dichloromethane solution and distill off the solvent. Upon standing, the heavy solution of methyl benzoate separates. It is a colorless liquid, distil from an air bath at 198-200 ^{0}C.

Q 1. What compound passes into the sodium bicarbonate solution?

4.23 Preparation of Soap

Fats and oils belong to the family of esters particularly the esters of glycerol and long chain carboxylic acids. Soaps are carboxylate salts with very long hydrocarbon chains. Soap can be made from the base hydrolysis of a fat or an oil. This hydrolysis is called saponification and the reaction has been known for centuries.

Traditionally, soaps were made from animal fat and base sodium hydroxide. An example of a saponification reaction is shown below.

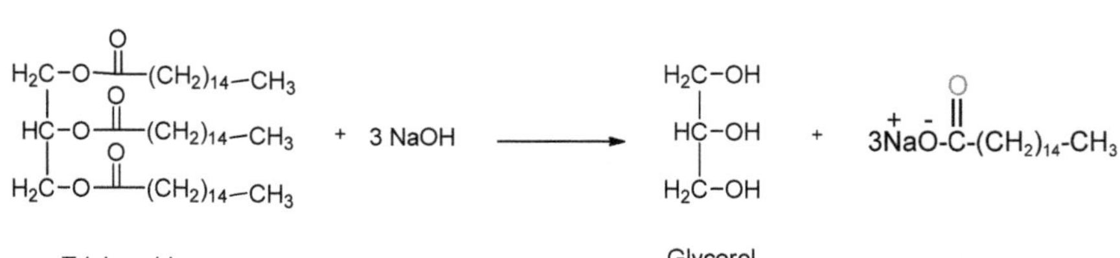

As you may remember, fats and oils are triesters of glycerol and three fatty acids. Esters can be hydrolyzed to their alcohol and carboxylic acid components in the presence of acid or base. Fats, oils, and fatty acids are insoluble in water because their hydrophobic tails are so long. If a base is used for hydrolysis, the fatty acids produced are deprotonated and are present as the corresponding carboxylate salts. Because these product carboxylate salts are charged, they are much more soluble in water than the corresponding uncharged fatty acids. Since the carboxylate salts also each have a long nonpolar tail, they are also compatible with nonpolar greases and oils.

Experimental procedure: Saponification (preparation of soap): In a 250 ml conical flask or a beaker weigh 20 g of cotton seed oil or any other fat, and add 12 ml of ethanol and 20 ml of 20 % NaOH to the beaker. Heat the mixture present in the beaker on a water-bath (with constant stirring) maintaining the bath temperature between 80-90° C. and continue heating for 1 hour or so, until the solution no longer has two separate layers. The solution should be transparent at this point. When saponification is complete, carefully remove the beaker from the water-bath. After this period add 150 ml of saturated sodium chloride solution and cool the mixture. This process is called "salting out". It decreases the solubility of the soap in the solution and therefore causes the soap to precipitate. Filter it through double the thickness of cheese cloth. Wash the soap on the cloth with 50 ml cold water and mould it into cake in a China dish or other suitable moulds.

To recover glycerol (a by-product) slightly acidify the filtrate obtained from above with HCl and evaporate it to dryness. Extract glycerol with 20 ml of absolute alcohol. Decant the alcohol solution from the salt and evaporate on a hot water bath.

4.24 Preparation of Benzocaine

Aim: To prepare an anaesthetic drug-benzocaine (Ethyl–*p*-aminobenzoate)

Principle: It consists of two steps.

(i) Reduction of p-nitro benzoic acid to *p*-amino benzoic acid.

(ii) Esterification of carboxylic acid to give benzocaine.

Step (i) Preparation of p-amino benzoic acid

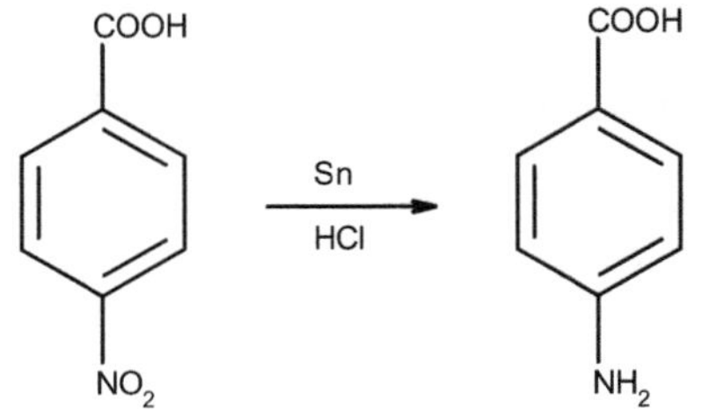

Experimental procedure: Place 2.5 g of p-nitro benzoic acid in 500 ml RB flask and to this add 6 g of tin granules, 12.5 ml of conc. hydrochloric acid. Fit with a reflux condenser heat the mixture gently until the reaction commences, and remove from flame. Shake the flask frequently and take care that the insoluble acid adhering to the side of the flask is transferred to the reaction mixture. Occasional gentle warming may be necessary. After about 20 min most of the tin would have reacted and a clear solution remains. Allow it to cool and decant the liquid into a 500 ml beaker. Wash the reminder tin by decantation with 8 ml of water and add the washings to the contents of the beaker. Add conc. ammonia solution until the solution is just alkaline to litmus. Filter off the precipitate of hydrated tin oxide and wash well with water. Filter off any solid which separates. Acidify the liquid with glacial acetic acid and evaporate on water bath until the crystal starts separating. Cool in ice filter the crystals at the pump. The yield of the product is 2 g, m.p. 192^0 C.

Step (ii) Preparation of ethyl *p*-amino benzoate

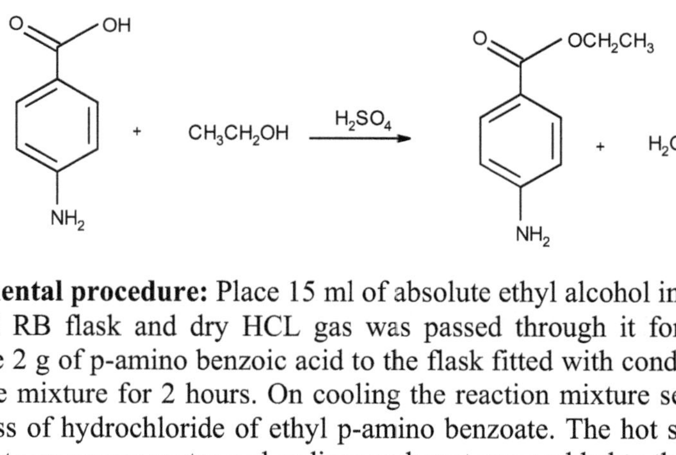

Experimental procedure: Place 15 ml of absolute ethyl alcohol in a 250 ml 2-necked RB flask and dry HCL gas was passed through it for 15 min. Introduce 2 g of p-amino benzoic acid to the flask fitted with condenser and reflux the mixture for 2 hours. On cooling the reaction mixture settled to a solid mass of hydrochloride of ethyl p-amino benzoate. The hot solution is poured into an excess water and sodium carbonate was added to the solution until it was neutral to litmus. The precipitate is filtered at the pump and dried in the air. A pure product may be obtained by recrystallisation from a small volume of ethanol. The yield of the material, is 1 g m.p. 91 ^{O}C.

Reference: Vogel's Textbook of Practical Organic Chemistry by Brain S, Furniss, Antony J. Hannaford, Peter W. G. Smith & Austin R. Tatchell: Fifth Edition.

CHAPTER 5

Spectroscopy

5.1 Ultraviolet Absorption Spectroscopy

The ultraviolet region falls in the range between 190-380 nm, the visible region falls between 380-750 nm.

Terminology for absorption shifts

(i) Chromophore: A covalently bonded unsaturated group which is responsible for absorption of UV – Visible radiations.

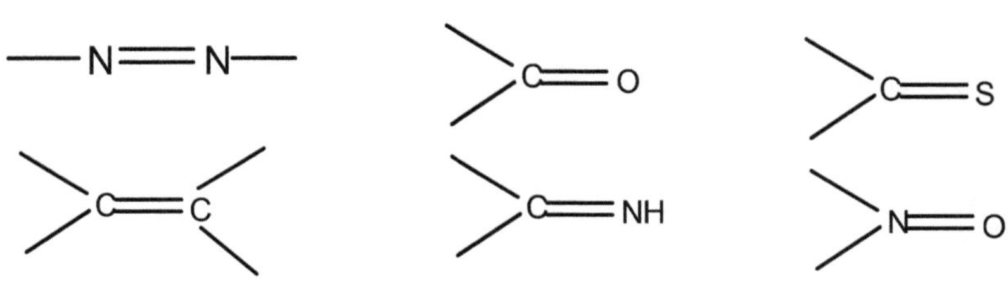

(ii) Auxochrome: A saturated group with 'n' electrons, which, when bonded to a chromophore, alters the wavelength & intensity of the absorption.

—OH —NHR —Cl —Br

$—CH_3$ $—NO_2$ $—NR_2$ —COOH

$—NH_2$ $—SO_2H$

(iii) Bathochromic Shift: The shift of absorption to a longer wavelength due to substitution or solvent effect (Red shift).

$H_3C - CH_2 - CH_2 - CH = CH_2$ $H_2C = C(CH_3) - CH = CH_2$

178 nm 222 nm

(iv) Hypsochromic Shift: The shift of absorption maxima, to shorter wavelength due to substitution or solvent effect (Blue shift)

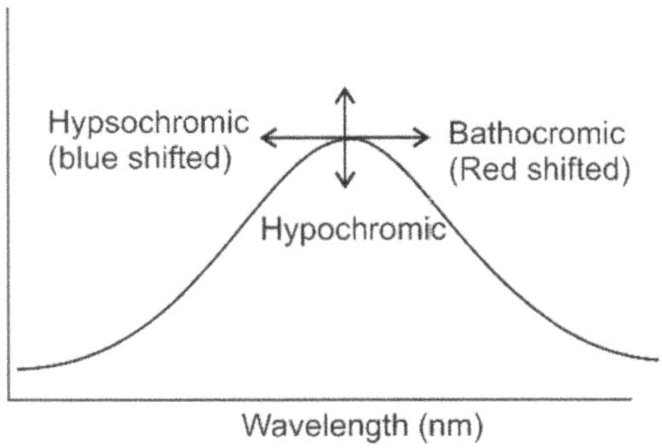

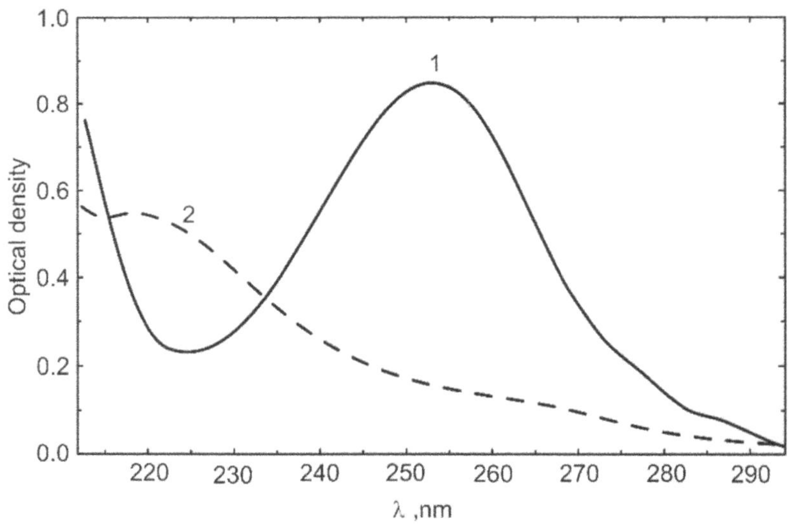

Fig.5.1 Absorption spectra of benzaldehyde
(Dotted line shows continuing spectrum beyond 295 nm).

Applications of Ultraviolet Spectroscopy

Alkanes, alkenes (non-conjugated), alkynes, alcohols, ethers, aliphatic carboxylic acids, aliphatic esters, aliphatic amides, nitriles: no absorption in U.V., no peak in 200 – 400 nm (i.e., transparent in U. V).

(a) Aliphatic aldehydes and ketones only **n** to **π*** transition

A weak peak in the region 275 – 295 nm ($\log \varepsilon = 1 - 2$)

(b) Compounds having sulfur and bromine ***n*** to **σ*** transition
A weak peak in the region 210 – 250 nm (log = 1 – 2)

(c) **Aromatic compounds: π** to **π***(B - band) transition
Medium intensity peak/s in the region 250 – 285 nm (log = 2)

(d) Conjugated dienes **π** to **π*** 220 – 230 nm (log = 4) transition

(e) **Enones:** 2 peaks in the region 220 -250 nm, intense peak due to **π** to **π*** transition and 300 to 350 nm weak **n** to **π*** transition

5.2 Infrared Spectroscopy (IR)

Theory

Infrared (IR) spectroscopy relies on the infrared region of the electromagnetic spectrum and is used to analyze solid, liquid and gaseous samples. The analysis of the samples is made possible by the fact that molecules absorb specific frequencies that are characteristic of their structure.

Passing a beam of infrared light through a sample and then measuring the absorption, emission or reflectance of the molecule records a spectrum. When a molecule is subjected to infrared radiation of a frequency corresponding to one of its natural vibration frequencies, then radiation is absorbed and that particular vibration excited. Different types of bonds vibrate at different frequencies, therefore, it is possible to derive information about the types of bonds in the molecule by observing the infrared absorption frequencies of a compound and hence deduce the nature of its functional groups. Most of the common organic functional groups absorb in the wave length range 2.5-1.5 um but the positions of absorption maxima are more commonly quoted in terms of frequencies, i.e. wavenumbers (v) and expressed in cm^{-1}.

An improvement of IR spectroscopy is the technique of Fourier transform infrared spectroscopy (FTIR). The infrared light is passed through the interferometer before being passed through a sample. The advantage of this technique is that all the frequencies are collected simultaneously and this in turn improves the signal-to-noise ratio as well as sampling speed. Nearly all infrared spectrometers these days are actually FTIR machines. FTIR spectroscopy has many applications, including in gas analysis, semiconductors, microelectronics and polymer manufacturing.

Procedure

When infrared light is passed through a sample of an organic compound, some of the frequencies are absorbed while other frequencies are transmitted through the sample without being absorbed. If we plot the % absorbance or % transmittance against frequency, the result is an infrared spectrum.

which contains the sample must not itself absorb infrared radiation. The most convenient material for the cell is sodium chloride in the form of polished rock salt plates, which are fragile and must be handled carefully. The samples therefore MUST be dry in order to avoid 'fogging' of the salt faces which reduces the amount of radiation transmitted.

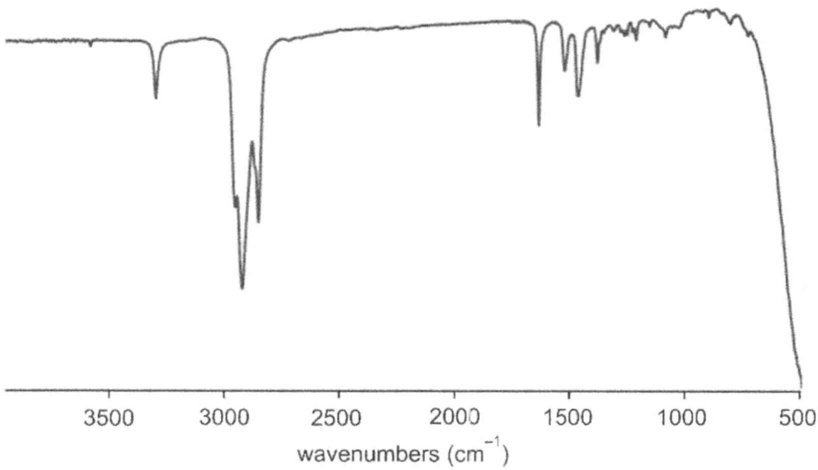

Fig.5.2 IR Spectrum in a Nujol mull.

Interpretation of Infrared Spectra

The region below 1500 cm^{-1}, which is often complex, is referred to as the fingerprint region. Confirmation of an unknown compound can thus be obtained by comparison of the infrared spectrum of the unknown with that of an authentic sample.

For the purpose of obtaining structural information from an infrared spectrum, it is convenient to divide the spectrum into 5 regions; the principal absorptions observed in these regions.

Region 1: 2850-4000 cm^{-1}. Principal absorptions in this area are due to single bonds involving hydrogen: C-H, O-H, N-H.C-H vibrations normally occur at the lower end of the range below 3100 cm-1. It may be possible in some cases to distinguish saturated C-H (ca 2900 cm^{-1}) from unsaturated C-H (ca. 3000-3100 cm^{-1}) bonds.

There are notable exceptions to this range: If the spectrum shows a strong absorption above ca. 3050 cm^{-1}, it almost certainly contains an O-H, or N-H or NH_2 group. There are three classes of O-H vibrations depending on the environment of the hydroxyl group. The tendency to form intermolecular hydrogen bonds increases as the acidity of the hydroxyl hydrogen increases, so the frequency of the O-H absorption decreases and the peak becomes broader. (N.B. If the sample has absorbed water it will show an O-H absorption).

N-H vibrations in both amines and amides show absorptions at 3150-3500 cm^{-1}.

Region 2: 1850-2850 cm^{-1}. Principal absorptions in this region are due to triple bonds, i.e. alkynes and nitriles typically appear at ca. 2100-2300 cm^{-1}.

Region 3: 1600-1850 cm^{-1}. This narrow range of frequencies is important because the principal absorption in this region are due to double bonds, i.e. C=C, C=N and C=O.

By far the most useful of these for the purpose of detecting functional groups is the carbonyl group absorption. This is most invariably a very strong absorption and occurs in a region 1775-1640 cm^{-1} where no other groups absorb with comparable intensity. For isolated (that is, non-conjugated) groups typical ranges are:

1755 ±20 esters

1730 ±15 aldehydes

1715 ±15 ketones

1710 ±15 carboxylic acids

1660 ± 30 amides

Some, but not all, aromatic compounds show a very sharp absorption around 1600 cm^{-1} (due to aromatic C=C).

Region 4: 1000-1600 cm^{-1} & Region **5**: 667-1000 cm^{-1}. These regions often contain a complete absorption pattern, only a few of these absorptions are of diagnostic importance.

Summary of diagnostic peaks in the Infrared Spectrum and functional group tests used to confirm identity.

Neutral compounds

Alcohols

(a) IR: peak at 3200-3600 cm^{-1}

(b) Form colored complexes with ceric (IV) ammonium nitrate [1]

(c) 1^0 and 2^0 alcohols reduce $Cr_2O_7^{2-}$ giving a green color [2].

Alkanals and Alkanones

(a) IR: peak at 1660-1720 cm^{-1}

(b) Form yellow, orange or red precipitate with 2,4-dinitophenylhydrazine (Brady's reagent)

(c) Alkanals reduce Tollen's reagent to form a silver mirror.

N.B. Ensure that the material to be tested is not basic, i.e. amine. Addition of amine to Brady's reagent may give a precipitate of 2,4-dinitrophenylhydrazine, or the sulfate of the amine. Both precipitates will re-dissolve in excess of reagent.

Primary Amides

(a) IR: peak at 1600 cm^{-1} (C=O) and two peaks at 3200-3300 cm^{-1} (NH_2)

(b) On heating with dilute sodium hydroxide, ammonia is evolved.

Acidic Compounds -- Carboxylic Acids

(a) IR: peak ca. 1700 cm^{-1} (C=O) and very broad absorption at 2500-3200 cm-1.

(b) Dissolve in sodium bicarbonate with evolution of CO_2 (not always easy to see).

Basic Compounds – Amines

(a) IR: Two peaks at 3200-3300 cm^{-1} (NH_2).

(b) **Nitrous acid test.** 1^0 aliphatic amines – Vigorous evolution of N_2. 1^0 aromatic amines – no apparent reaction but will form a red azo dye when an alkaline solution of 2-naphthol is added. **N.B**. Some evolution of gas will always occur.

1. **Ceric (IV) Ammonium Nitrate Test:** Dissolve 1 drop or 20 mg (if solid) of a given compound in water (or dioxane it is insoluble in water). Add to 0.5 ml of cerium (IV) reagent; orange or red color indicates alcohol is present.

2. **Oxidation with Dichromate Test:** Add 1-2 drops or 20 mg (if solid) of the given compound to a solution of acid dichromate (1 ml). Green color indicates reduction of orange Cr (VI) to green Cr (III) and the presence of an oxidisable group.

5.3 Nuclear Magnetic Resonance (NMR)

Over the last 50 years, nuclear magnetic resonance (NMR) spectroscopy has become a widely used and effective analytical chemistry technique. So far there have been six Nobel prizes awarded to researchers connected to the technique.

NMR relies on the fact that nuclei in a magnetic field absorb and re-emit electromagnetic radiation. NMR spectroscopy takes advantage of the magnetic properties of certain atomic nuclei and can be used to determine information about the structure, dynamics, reaction state and chemical environment of molecules.

Most NMR instruments require the samples to be in solution, but the development of solid-state NMR has allowed the analysis of crystals, powders and molecular aggregates. High- resolution data can be gathered using the magic-angle spinning and cross polarization techniques.

General notes on Presentation of Spectra

By convention all field-sweep NMR spectra are plotted with the direction of increasing magnetic field going from left to right. In the case of frequency-sweep spectra the equivalent presentation is obtained with direction of decreasing frequency going from left to right.

The direction from left to right is referred to (in the case of both field- and frequency – sweep) as the **upfield** direction, the opposite direction is referred to as the **downfield** direction.

A nucleus giving rise to a signal further upfield than another nucleus is said to be more **shielded**. A nucleus giving rise to a signal further downfield than another is said to be **de-shielded**.

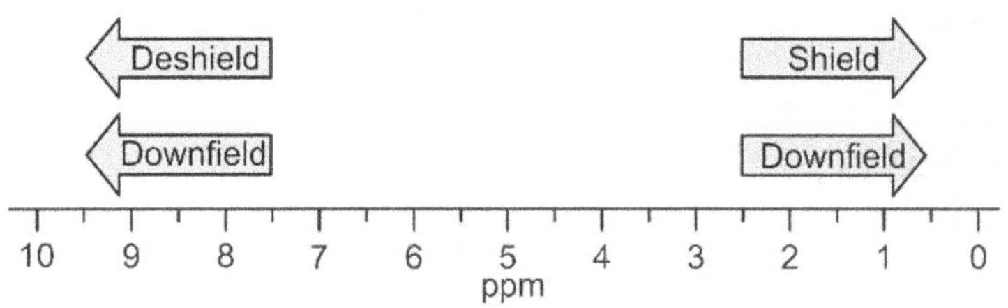

Fig. 5.3

Movement of a resonance signal upfield is often referred to as a **diamagnetic** shift, while movement in the downfield direction is known as a **paramagnetic** shift.

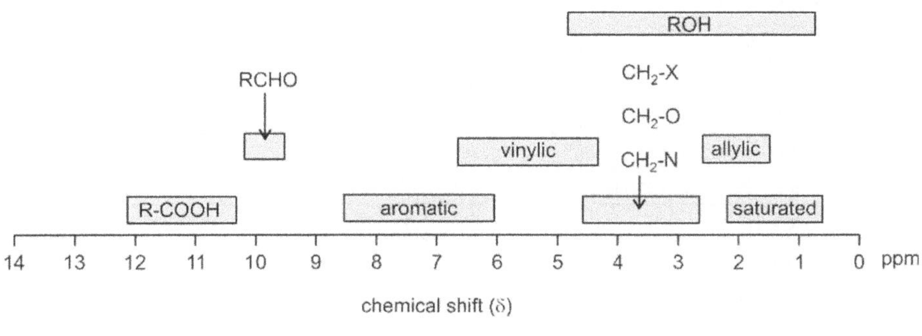

Fig. 5.4 ^{1}H chemical shift ranges for organic compounds.

Aromatic Compounds

In aromatic compounds **π** - electrons are cyclically delocalized over the aromatic ring. When these compounds are kept in an external magnetic field, H_0–these closed loop of **π** - electrons are induced to rotate further.

Shielding: If the induced magnetic field opposes the applied magnetic field at the proton, then the proton is said to be shielded and the phenomenon is known as shielding.

De-shielding: If the induced magnetic field aligns with the applied magnetic field at the present, then the proton is said to be deshielded and the phenomenon is known as de-shielding.

In aromatic rings the circulation of electrons produce a substantial electric current called the ring current and of course an associated magnetic field. This field in the centre of the ring opposes the applied external field (diamagnetic effect) but the field around the outside of the ring enhances the instrument field and the aromatic protons are deshielded (see below). Aromatic protons therefore appear in the spectrum at around (see chemical shift tables).

Typical Chemical Shifts of Hydrogen (where R = alkyl &Ar = aryl)

Groups	(δ)	(τ)
RCH_3	0.9	9.1
$R2CH_3$	1.3	8.7
R_3CH	1.5	8.5
R-CHO	9.5 -10.0	0.5 – 0
$H_2C = CH_2$	4.0 - 8.0	6.0 - 2.0
HC≡CH	1.5 - 3.5	8.5 – 6.5
ROH	0.5 – 4.5	9.5 – 5.5
ArOH	4.5	5.5
R-COOH	10.5 – 13.0	-0.5 to -3.0
Ar-CHO	10.1	-0.1
$Ar\text{-}OCH_3$	3.3 – 4.0	6.0 – 6.7
Ar-H	6.0 – 8.5	4.0 – 1.5
RNH_2, $RNHR^1$	5.0 – 8.0	5.0 – 2.0
$ArNH_2$, ArNHR	3.5 – 6.0	6.5 – 4.0
$RCONH_2$, RCONHR	5.0 – 8.5	5.0 – 1.5
RSH	1.0 – 2.0	9.0 – 8.0
ArSH	3.0 – 4.0	7.0 – 6.0
=NOH	10.0 – 12.0	0 to -2 (often broaden peak)
OC – H (esters)	3.7 – 4.1	6.3 – 5.9

Table *contd...*

Groups	(δ)	(τ)
CH_3I	2.16	7.84
CH_3Br	2.65	7.35
CH_3CL	3.10	6.90
CH_3F	4.26	5.74

***The spectrum of ethanol is shown below*:**

Compound ethanol contains three types of protons and they absorb at different **δ** values and correspond to OH group (δ 4.65), $-CH_2$ (δ 3.55) and $-CH_3$ (δ1.2)**.** The CH_3 group is a triplet because of coupling to CH_2. The OH signal is a triplet because of coupling to CH_2. The CH_2 is split into a quartet by the neighboring CH_3 group. The area beneath each peak corresponds to a numerical ratio of 1:2:3.

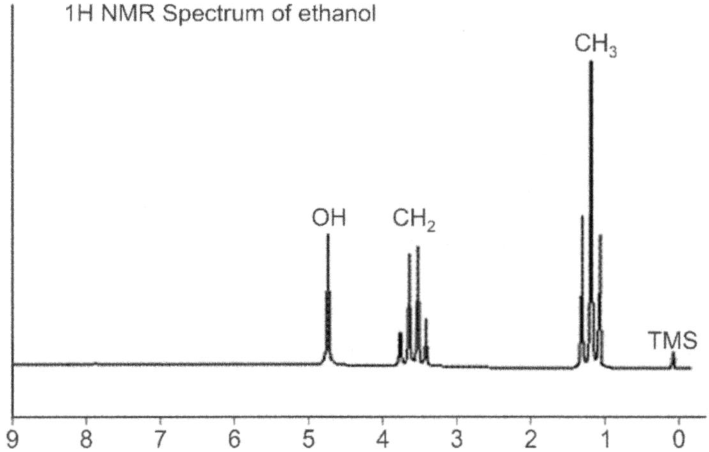

Fig. 5.5

5.4 Mass Spectrometry

Mass spectrometry is not an absorption spectroscopy unlike UV, IR and NMR spectroscopy.

It is a destructive method. Therefore, the compound cannot be recovered.

Fortunately, we require only very small quantity of the compound (often less than 1 mg) to record a mass spectra.

Mass spectrometry is useful for

(a) Determining the molecular weight of the compound
(b) Determining the molecular formula
(c) Detection of S, N, Cl, Br, I, in a molecule

(d) Detection of certain groups in organic compounds like $-CH_3$; $-CH_2-CH_3$; $-CH_2-CH_2-CH_3$; -OH, $—NH_2$, —CN, —Ph, Ph - CO etc.

Basic Principle

In the mass spectrometer, organic molecules are bombarded with a high energy electron beam. This results in the loss of one loosely bound electron from the molecule and forms a highly energetic positively charged ions called molecular ions or parent ions.

These molecular ions, further break & produce smaller ions called daughter ions which give a characteristic ''fingerprint'' of the molecule.

IR spectrum

Some of the more common fragments which appear as ions and /or lost as radicals or molecules in mass spectra.

m/e	Fragment	m/e	Fragment
1	H	2	H_2
14	CH_2	16	O, NH_2
17	OH	18	H_2O, NH_4
19	F	20	HF
26	C_2H_2, CN	27	C_2H_3, HCN
28	C_2H_4, CO, N_2	29	C_2H_5, CHO
30	CH_2NH_2, NO	31	CH_2OH, OCH_3
32	CH_3OH	33	SH
34	H_2S	35	Cl
36	HCL	41	C_3H_5
42	C_3H_6	43	C_3H_7, $COCH_3$
44	CO_2, $C_2H_4NH_2$	45	C_2H_5O, COOH
46	NO_2	51	C_4H_3
55	C_4H_7	57	C_4H_9, C_2H_5CO
59	$COOCH_3$	71	C_5H_{11}
77	C_6H_5	78	C_6H_6
80	HBr	91	C_7H_7
105	COC_6H_5	127	I
128	HI		

Annexures

Annexure - I

Solutions

Types of Solutions

(a) Molar (M): A molar solution is one containing one mole or gram-molecular weight of the solute in 1 litre of solution. The term molar is used for a solution containing 1 mole in 1000 grams of solvent.

(b) Normal (N): A normal solution contains one gram-equivalent weight of solute in 1 litre of solution.

N = Weight of solute per litre of solution/Equivalent mass

Relation between N and M

$$N = M \times n$$

Where n = number of e- change

N = Normality; M = Molarity

(c) Isotonic: Isotonic solutions are those that exert equal osmatic pressures. Isotonic saline is 8.5 g NaCl in 1 litre of aqueous solution.

(d) Aqueous: Aqueous solutions are, of course, those using water as solvent. Where no solvent has been mentioned it can be assumed to be water.

(e) Percentage solution: Percentage solution of solid solution refer to grams per 100 ml of solution, percentage mixtures of liquids refer to ml of one to 100 ml of mixture (example 60% of ethanol is 60 parts of ethanol plus 40 parts of water).

(f) Weight/volume and volume/volume: The abbreviations w/v and v/v are sometimes added after the strength of a solution. Thus, trichloroacetic acid 15% w/v is a solution containing 15g of trichloroacetic acid per 100 ml of solution, and 75% ethanol v/v is a solution containing 75 ml of ethanol per 100 ml of solution.

(g) Buffer solution: Buffer solution that tend to resist change in pH on the addition of small quantities of acid or alkali. Many buffer solutions are mixtures of either a weak acid and one of its salts or a weak base and one of its salts. Commonly used buffer systems are:

(i) **Phosphate:** Potassium hydrogen orthophosphate (KH_2PO_4) + disodium hydrogen orthophosphate (Na_2HPO_4).

(ii) **Acetate:** Acetic acid + Sodium acetate

(iii) **Citrate:** Citric acid + Sodium citrate

(iv) **Barbitone (Veronal):** Sodium diethylbarbiturate + HCl

Dilution

Diluting from one strength to another. A simple rule to remember when diluting from one strength to another is as follows: If **S** is the strength of the given solution and **t** is the required strength, take **t** parts of the given solution and make up to **S** parts with solvent. For example, a 10% aqueous solution may be obtained from a 25% solution by taking 10 ml of the 25% solution and making upto 25 ml with water.

If a definite volume of dilution is needed use the following formula.

$$tV/S = Vs$$

Where

V = Volume required,

Vs = Volume of solution of concentration **S** to be diluted.

Example: Given a 36% solution, prepare 500 ml of a 9% solution.

Substitution in the formula above

$$9 \times 500/36 = Vs = 125$$

Therefore take 125 ml of the 36% solution and dilute to 500 ml.

Acids and Bases

C.C. per litre to make approx					
		Acids and Bases			
		Approx Normality	**1.0N (N/1)**	**0.1N (N/10)**	**0.01N (N/100)**
Acids:	S.G	N	CC	CC	CC
Glacial acetic acid	1.06 (99%)	16	57	5.7	0.6
Hydrochloric	1.16 (36%)	12	86	10	1.0
Nitric	1.42 (69%)	16	63	6.3	0.7
Sulphuric	1.84 (95%)	36	28	2.8	0.3
Bases:					
Ammonium hydroxide	0.88	20	50	5.0	0.5
Potassium hydroxide	Solid	Solid	58 g	5.8 g	0.6g
Sodium hydroxide	Solid	Solid	42 g	4.2 g	0.42 g

Approximate Normality

Dilute acids: Use the amount of concentrated acids or bases as mentioned and dilute to one litre to make appropriate normality.

(i) **Dilute acids for general use**

(a) **Hydrochloric acid (3N):** Add 258 cc of conc. hydrochloric acid to water and dilute to 1 litre.

(b) **Nitric acid (3N):** Add carefully 189 cc of conc. nitric acid, by external cooling, to water and make up the volume to 1 litre.

(c) **Sulphuric acid (3N):** Add slowly 84 cc of con. sulphuric acid, by external cooling, to cold water and make upto the volume to 1 litre.

(d) **Acetic acid (3N):** Add 172 cc of glacial acetic acid to water and dilute to 1 litre.

(e) **Dilute hydrochloric acid (5% v/v, 1.4N):** Dilute 5 cc of conc. HCl with water until total volume is 100 cc.

(f) **Dilute acetic acid (5% v/v, 0.8N):** Dilute 5 ml of glacial acetic acid with water until total volume is 100 cc.

(g) **Dilute sulphuric acid (5% v/v, 1N):** Add 5 cc of conc. H_2SO_4 to cold water and make upto 100 cc with water.

(h) **Dilute nitric acid (5% v/v, 0.8N):** Add 5 cc of conc. HNO_3 carefully to water by external cooling, and make up to 100 cc with water.

(ii) **Dilute bases for general use**

(a) **Sodium hydroxide (10% aqueous, 2.5N):** Dissolve 10 g of NaOH in water and dilute to 100 ml.

(b) **Sodium hydroxide (30% aqueous, 7.5N):** As above, using 30 g of NaOH.

(c) **Ammonium hydroxide (dilute):** Dilute 1 volume of conc. ammonia (d, 0.88) to 3 vols. of water.

(d) **Potassium hydroxide (3N):** Dissolve 176 g of the pellets in water and dilute to 1 litre.

Indicators

(a) **Litmus (Indicator) solution:** Extract litmus powder three times with boiling alcohol, each treatment consuming an hour. Discard or reject the alcohol extract. Treat residue with an equal amount of cold water and filter; then extract with five times its weight of boiling water, cool and filter. Combine the aqueous extracts.

(b) **Methyl orange:** Dissolve 0.1 g of methyl orange in 100 ml of water filter, if necessary.

(c) **Methyl red:** (i) Dissolve 0.1 g of methyl red in 60 ml of alcohol and dilute with 40 ml of water. (ii) Methyl red modified: Dissolve 120 mg of methyl red and 80 mg of methylene blue in 100 cc of 95% alcohol.

(d) **Phenolphthalein:** Dissolve 1 g of phenolphthalein in 50 ml of alcohol and add 50 ml of water.

(e) **Thymol-blue:** Dissolve 0.5 g of thymol-blue in 100 ml of ethanol and add 200 ml of water.

Annexure - II

Laboratory Reagents

1. **Ammonium acetate (1N):** Dissolve 73 g of ammonium acetate ($NH_4C_2H_3O_2$) in water and dilute to 1 liter.
2. **Ammonium carbonate (1N):** Dissolve 96 g of the analar grade salt ($(NH_4)_2CO_3.H_2O$) (1N) in 170 ml of 1N NH_4OH and dilute to 1 liter.
3. **Ammonium chloride (1N):** Dissolve 53.5 g of NH_4Cl in 1 liter of water.
4. **Ammonium nitrate (1N):** Dissolve 80 g of NH_4NO_3 in 1 liter of water.
5. **Ammonium oxalate (0.5N):** Dissolve 35.5 g of ammonium oxalate ($(NH_4)_2C_2O_4.H_2O$) in water. Dilute to 1 liter.
6. **Ammonium sulphate (0.5 N):** Dissolve 33 g $(NH_4)_2SO_4$ in 1 liter of water.
7. **Ammonium sulphide, colourless (1N):** Treat 68 ml of conc. (NH_4OH) ammonium hydroxide solution with H_2S until saturated, keeping the solution cold. Add 68 ml of conc. NH_4OH and dilute to 1 liter.
8. **Ammonium molybdate (0.5 N):** Dissolve 45 g of pure ammonium molybdate in a mixture of 40 ml of aqueous ammonia (d 0.88) and 60 ml of water and then adding 120 g of ammonium nitrate and dilute the solution to 1 liter with water.
9. **Ammonium thiocyanate (1N):** Dissolve 76g of ammonium thiocyanate (NH_4CNS) in water and dilute to 1 liter.
10. **Ammonium vanadate:** Dissolve 60 mg of ammonium vanadate in 200 ml of water.
11. **Ammoniacal silver nitrate solution (Tollen's reagent):** Tollen's reagent is unstable and must be prepared just prior to use. Dissolve 5 g

of silver nitrate in 100 ml of water (**solution 1**) and 5 g of sodium hydroxide in 100 ml of water (**solution 2**). Keep these two solutions separately until required. When the reagent required, mix equal volumes (say, 1 ml each) of **solution 1** and **solution 2** in a clean test tube, and then add dilute ammonia solution (concentrated ammonium hydroxide diluted with 5 volumes of water) drop by drop with shaking until the silver oxide is just dissolved. Only small volume should be prepared just before use and any residue is to be washed down the sink with a large amount of water (Note: An excess ammonia will decrease the sensitivity of the reagent). Great precautions must be taken in the preparation and use of this reagent, **which must not be heated**.

12. **Barium chloride (0.25M, 0.5N):** Dissolve 61 g of barium chloride ($BaCl_2.2H_2O$) in water. Dilute to 1 liter.

13. **Benedict's reagent (Test for reducing substances):** Dissolve 43 g crystalline sodium citrate and 25 g of anhydrous sodium carbonate in 175 ml of water. Filter if necessary and add a solution of 4.32 g of crystalline copper sulphate in 25 ml of water with stirring. Make up to 250 ml water.

13a **Barfoed's reagent for reducing mono saccharides:** Dissolve 6.6 g of cupric acetate and 1 ml of glacial acetic acid in water and dilute to 100 ml.

14. **Calcium chloride (0.25M, 0.5N):** Dissolve 55 g of $CaCl_2.6H_2O$ in water. Dilute to 1 liter.

15. **Chlorine-water:** Pass chlorine gas ($KMnO_4$ + Conc. HCl) into cold water till a saturated solution is obtained and keep it in a dark brown bottle.

16. **Ceric ammonium nitrate:** Dissolve 20 g of ceric ammonium nitrate in 500 ml warm dilute (2 N) nitric acid.

17. **Dichromate solution:** Dissolve 20 g of potassium or sodium dichromate in a mixture of 150 ml of water and 50 ml conc. sulphuric acid.

18. **2,4-Dinitrophenylhydrazine (2,4-DNP):** Dissolve 2 g of 2,4-DNP in 15 ml conc. sulphuric acid added cautiously to 500 ml of 30 percent alcohol.

19. **Disodium hydrogen phosphate (1N):** Dissolve 60 g disodium hydrogen phosphate ($Na_2HPO_4.12H_2O$) in water. Dilute to 500 ml.

20. **Ferric chloride (0.5M, 1.5N):** Dissolve 135 g of ($FeCl_3$) ferric chloride in water containing 20 ml of conc. HCl dilute to 1 liter.

21. **Ferrous sulfate (0.5M, 1N):** Dissolve 13.9 g of ferrous sulphate ($FeSO_4.7H_2O$) in water containing 1.0 ml of conc. sulphuric acid. Dilute to 100 ml.

22. **Fehling's solution, Solution No.1:** Dissolve 34.6 g of copper (II) sulphate crystals in water containing a few drops of dilute sulphuric acid, and dilute the solution to 500 ml.

    Fehling's solution, Solution No.2: Dissolve, while cooling, 60 g of pure sodium hydroxide and 173 g of pure sodium potassium tartrate (Rochelle salt) in water, filter if necessary through a sintered glass funnel and make up the filtrate and washings to 500 ml.

    Keep the two solutions separately in tightly stoppered bottles and mix exactly equal volumes (say 1 ml each) immediately before use.

23. **Ferrous ammonium sulphate (Mohr's salt):** Dissolve 19.6 g of $(FeNH_4SO_4)_2.6H_2O$ in water containing 1 ml of conc. H_2SO_4. Dilute to 100 ml. prepare fresh solutions for best results.

24. **Formaldehyde-sulfuric acid (Marquis' reagent for alkaloids):** Add carefully 10 ml of formaldehyde solution to 50 ml of sulphuric acid.

25. **Iodine solution:** Add 2 g of powdered iodine to a solution of 4 g of potassium iodide in a minimum amount of water and make upto 300 ml with water.

26. **Lead acetate (0.5M, 1N):** Dissolve 19.0 g of lead acetate in water and dilute to 100 ml.

27. **Lead nitrate (0.25M, 0.5N):** Dissolve 8.3 g of $Pb(NO_3)_2$ in water. Dilute to 100 ml.

28. **Lime water (Calcium hydroxide), (0.02M, 0.04N):** Dissolve 1.5 g of calcium hydroxide or calcium oxide in water and make up to 1 liter.

29. **Lucas reagent:** To concentrated hydrochloric acid (5 ml) add anhydrous zinc chloride (2 g) and boil for 1 minute, cool to room temperature.

30. **Mercuric chloride (0.1N):** Dissolve 1.35 g mercuric chloride ($HgCl_2$) in water and dilute to 100 ml water.

31. **Molish's reagent (General test for carbohydrate):** Dissolve 10 g of 2-naphthol in 100 ml ethanol (95%).

32. **Nessler's reagent (for ammonia):** Dissolve 5 g of KI in the smallest quantity of cold water (5 ml). Add a saturated solution of mercuric

chloride (about 2.2 g in 35 ml of water) until an excess is indicated by the formation of a precipitate. Then add 20 ml of 5 N NaOH and dilute to 100 ml. Let settle, and draw off the clear liquid.

33. **Neutral Ferric chloride:** Prepare the neutral ferric chloride solution (i.e., free from HCl) by adding dilute sodium hydroxide solution to the bench ferric chloride reagent until a slight precipitate of iron(III) oxide is formed. Filter off the precipitate and use the clear solution.
34. **Ninhydrin solution:** Dissolve 0.2 g ninhydrin in 100 ml distilled water.
35. **Potassium dichromate (0.125M):** Dissolve 37 g potassium dichromate ($K_2Cr_2O_7$) in 1 liter of water.
36. **Potassium ferricyanide (0.2N):** Dissolve 2.1 g of solid ($K_3Fe(CN)_6.3H_2O$) in water and dilute to 1 liter.
37. **Potassium Iodide-iodine solution (0.5M, 0.5N):** It is prepared by dissolving 20 g of potassium iodide and 10 g of iodine in 100 ml of water.
38. **Potassium hydroxide (alcoholic) solution:** A mixture of powdered potassium hydroxide (10 g) and rectified spirit (100 ml) is refluxed for 30 minutes, cool and decant through a filter, if solid material remains.
39. **Potassium permanganate solution (0.02M):** Dissolve 3.16 g of $KMnO_4$ in 1 litre water.
40. **Picric acid:** Prepare a saturated solution of picric acid in ethanol or benzene and filter the clear solution from undissolved portion of the solid.
41. **Potassium iodide mercuric iodide (reagent for proteins):** Dissolve 5 g of KI in 50 ml of water, and saturated with mercuric iodide (about 12 g). Dilute to 100 ml.
42. **Potassium ferrocyanide (0.5M, 2N):** Dissolve 21.1 g of potassium ferrocyanide in water. Dilute to 1 litre.
43. **Potassium carbonate (0.75M, 1.5N):** Dissolve 20.7 g of K_2CO_3 in 100 ml of water.
44. **Potassium chloride (0.5M, 0.5N):** Dissolve 37 g of KCl in 1 litre of water.
45. **Sodium carbonate (0.75M, 1.5N):** Dissolve 79.5 g of Na_2CO_3 in water. Dilute to 1 litre.
46. **Sodium chloride (0.5M, 0.5N):** Dissolve 29 g of NaCl in 1 litre of water.

47. **Sodium hydrogen phosphate (0.167M, 0.5N):** Dissolve 60 g of sodium hydrogen phosphate ($Na_2HPO_4.12H_2O$) in 1 litre of water.

48. **Sodium nitrate (0.05M, 0.05N)**: Dissolve 4.3 g of $NaNO_3$ in 100 ml of water.

49. **Sodium sulfide (0.05M, 0.5N):** Dissolve 60 g of solid $Na_2S.9H_2O$ in water and dilute to 500 ml of water.

50. **Stannous chloride (0.05M, 0.1N):** Dissolve 11.3 g of $SnCl_2.2H_2O$ in 17 ml of conc. HCl, using heat if necessary. Dilute with water to 100 ml.

51. **Sodium hypobromite:** Dissolve 100 g of NaOH in water by external cooling in ice-water and make upto 1 litre, add slowly 25 ml of bromine with stirring the mixture.

52. **Sodium hydrogen sulphite solution:** Dissolve 300 g of $NaHSO_3$ in water, make up to 500 ml and pass sulphur dioxide for a few minutes to remove sodium sulphite.

53. **Starch solution:** A solution of starch which keeps indefinitely is made as follows: mix 500 ml of saturated NaCl solution (filtered), 80 ml of glacial acetic acid, 20 ml of water and 3 g of starch. Bring slowly to boil and boil for two more minutes and filter if necessary; store in a glass stoppered bottle.

OR

Make a paste of 4 g of soluble starch in 5 ml of cold water. In a separate beaker boil 150 ml water and add cold paste till a clear solution is obtained.

54. **Sodium bicarbonate (10%) solution:** Dissolve 10 g of sodium bicarbonate in 100 ml of water.

55. **Sodium nitroprusside:** Dissolve 1 g of sodium nitroprusside in 100 ml of distilled water.

56. **Silver nitrate (0.05N):** Dissolve 8.5 g of $AgNO_3$ in water dilute to 100 l.

57. **Sodium acetate (1N) solution:** Dissolve 136 g of sodium acetate in water and dilute to 1 liter.

58. **Schiff's reagent:** Add one gram of sodium metabisulphite to a solution of 0.1 g of p-rosaline hydrochloride and 2 ml of conc. HCl in 100 ml of water.

59. **Tollen's reagent:** See ammoniacal silver nitrate solution.

Annexure - III

Important Solvents, Solids, and Liquid Reagents required in the Laboratory

One hundred gram quantities of the following solids and liquid reagents, used mainly in the preparation of derivatives.

Important Solvents, Solids and Liquid reagents required in the Laboratory

S.No	Chemical name	Chemical formula	Molecular mass (g/mol) atomic wt
1	Aluminum metal	Al	23.98
2	Ammonium chloride	NH_4Cl	53.49
3	Ammonium dichromate	$(NH_4)_2Cr_2O_7$	252.07
4	Ammonium hydroxide	NH_4OH	35.04
5	Acetic acid	CH_3COOH	60.05
6	Acetic anhydride	$(CH_3CO)_2O$	102
7	Acetyl chloride	CH_3COCl	78.5
8	Aniline ammonium chloride	$C_6H_5NH_3Cl$	129.5
9	Barium chloride	$BaCl_2$	208.23
10	Benzoyl chloride	C_6H_5COCl	140.5
11	Blue litmus paper	----	

Table *contd...*

12	Calcium chloride (dry)	$CaCl_2$	111
	p-Bromophenacyl bromide	$BrC_6H_4COCH_2Br$	278
	Copper metal	Cu	63.546
	Copper sulphate	Cu_2SO_4	223.15
	Chloroacetic acid	$ClCH_2COOH$	94.35
	Chlorosulfonic acid	$Cl.SO_3H$	116.45
	Diethylene glycol	$C_5H_{14}O_3$	134
	3,5-Dinitrobenzoic acid	$C_7H_4N_2O_6$	212
	2,4-Dinitochlorobenzene	$C_6H_3N_2O_4Cl$	202.45
	3,5-Dinitrobenzoyl chloride	$C_7H_3N_2O_5Cl$	94
	2,4-Dinitophenyhydrazine	2,4-$(O_2N)_2C_6H_3.NH-NH_2$	184
	Ferrous sulphate	$FeSO_4.7H_2O$	277.8
	Formic acid	HCOOH	46
	Glucose	$C_6H_{12}O_6$	180.16
	Hydrochloric acid	HCl	36.46
	Hydroxylamine hydrochloride	$NH_2OH.HCl$	69.45
	Iron fillings	Fe	55.845
	Iodine	I	126.9
	Lead nitrate	$Pb(NO_3)_2$	331.2
	Magnesium metal	Mg	24.305
	Magnesium sulphate	MgSO4	120
	Medicinal paraffin		
	Methyl iodide	CH_3I	142
	Methyl orange	$C_{14}H_{14}N_3NaO_3S$	327.33

Table *contd…*

	2-Naphthol	$C_{10}H_8O$	144
	Naphthalene	$C_{10}H_8$	128
	alfa-Naphthylamine	$C_{10}H_9N$	143
	p-Nitrobenzoyl chloride	$C_7H_4NO_3Cl$	185.4
	p-Nitrobenzyl bromide	$C_7H_6NO_2Br$	216
	Potassium chromate	K_2CrO_4	194.19
	Potassium dichromate	$K_2Cr_2O_7$	294.185
	Potassium iodide	KI	166
	Potassium nitrate	KNO_3	101.1
	Potassium permanganate	$KMnO_4$	158.03
	Potassium pentachloride	KCl_5	177.25
	Picric acid	$C_6H_3N_3O_7$	229
	Phophorus trichloride	PCl_3	106.35
	Phthalic anhydride	$C_8H_4O_3$	148
	Potassium hydroxide	KOH	56
	Pyridine	C_5H_5N	79
	Phenolphthalein	$C_{20}H_{14}O_4$	318.33
	Sodium bicarbonate	$NaHCO_3$	84
	Sodium carbonate	Na_2CO_3	105.98
	Sodium chloride	NaCl	58.44
	Sodium hydroxide	NaOH	39.99
	Sodium sulphate	Na_2SO_4	142.04
	Sodium thiosulphate	$Na_2S_2O_3$	158.11
	Sulphur	S	32.06
	Sodium acetate	$C_2H_3O_2Na$	82
	Semicarbazide hydrchloride	CH_5N_3O * HCl	111.45
	Sodium nitrite	$NaNO_2$	69

Table *contd...*

	Sodium nitroprusside	$Na_2[Fe(CN)_5NO]$	297.95
	Sodium hypochlorite	NaOCl	74.45
	Sodium metal under oil	Na	23
	Thionyl chloride	$SOCl_2$	118.9
	Thiourea	CH_4N_2S	76
	p-Toluene sulfonic acid	$C_7H_8O_3S$	172
	p-Toluene sulphonyl chloride	$C_7H_7SO_2Cl$	190
	p-Toluidine	C_7H_9N	107
	1,2,3-Trinitobenzene	$C_6H_3N_3O_6$	213
	2,4,6 – Trinitrotoluene	$C_7H_5N_3O_6$	227
	Tin granules	Sn	118.69
	Zinc sulphate	$ZnSO_4$	161.38
	Zinc metal	Zn	65.38
	High vaccume grease	------------	----
IMPORTANT SOLVENTS			
1.	Acetone	CH_3COCH_3	58
2.	Benzene	C_6H_6	78
3.	Carbon tetrachloride	CCl_4	153.80
4.	Chloroform	$CHCl_3$	119.35
5	Diethy ether	$C_2H_5OC_2H_5$	74
6.	Ethanol	C_2H_5OH	46
7.	Ethyl acetate	$CH_3COOC_2H_5$	88
8.	Hexane	C_6H_{14}	86
9	Methanol	CH_3OH	32
10	Petroleum ether	$C_6 - C_8$ carbons (appro)	

Annexure - IV

(a) Organic Solvents –Risk Assessment

Compound	Hazard	Spillage	Miscible with water	Disposal
Acetone	Flammable irritant	Absorb or Wash away	Yes	Waste solvents or down drain
Acetic Acid (glacial)	Flammable/Corrosive	Neutralize with Na_2CO_3 or wash away	Yes	Down drain
Dichloromethane	Toxic/Flammable	Absorb	No	Waste solvents
Ethanol	Flammable	Absorb/Wash away	Yes	Down drain
Ether	Flammable	Absorb	No	Waste solvents
Ethyl acetate	Flammable/Toxic	Absorb	No	Waste solvents
Methanol	Flammable/Toxic	Absorb	Yes	Down drain
Petroleum ether	Flammable	Absorb	No	Waste solvents
Toluene	Flammable/Irritant	Absorb	No	Waste solvents

(b) Inorganic Reagents – Risk Assessment

Compound	Hazard	Spillage	Disposal
HCl (aq.)	Corrosive/Toxic	Neutralize and wash away	Down drain
HNO_3	Corrosive/Oxidizing agent	Neutralize and wash away	Down drain
H_2SO_4	Corrosive/Oxidizing agent	Neutralize and wash away	Down drain
NH_4OH	Irritant/corrosive	Neutralize and wash away	Down drain
NaOH	Irritant/corrosive	Neutralize and wash away	Down drain
$CaCl_2$	Irritant	Wash away	Down drain
$FeCl_3$	Corrosive	Wash away	
$TiCl_3$	Corrosive	Wash away	

Annexure - V

Safety and Chemical Safety Data

In compliance with safety and health at work act and the Control of Substances Hazardous to Health Regulations, the following instructions, risk assessments and chemical hazardous data should be read in conjunction with the experiments described in this book.

The object of this laboratory course book is to develop your experimental skills and to teach you how to handle organic and other chemicals using safe procedures.

Safety glasses and laboratory coats must be worn at all times in the laboratory.

Students working in the organic chemistry teaching laboratory must follow safe procedures, with proper regard at all times for the wellbeing of all persons in the laboratory.

Before commencing an experiment, familiarize yourself with the hazardous properties of all materials to be used. Read the accompanying risk assessment.

Remember that nearly all organic compounds are flammable. Extreme caution should be exercised when working with volatile substances. Organic solvents must NOT be heated with a Bunsen burner. In the interest of safety, Bunsen burners may be used only in fume cupboards, but

(i) Do not light a burner in a fume cupboard already in use for another experiment.

(ii) Extinguish the burner immediately after use.

Find out the location of all the fire extinguishers, fire blankets or emergency showers.

Immediately clean up accidental spillages of all chemicals. Try to avoid skin contact with chemicals, and use disposable gloves when indicated, especially when using phenols. It is usually sufficient to remove chemicals from the skin by washing with soap and water.

Excess reagents and by-products should be disposed of in an appropriate manner; aqueous reagents should be neutralized and discarded in the fume-cupboard sinks with sufficient water.

Organic solvents should be placed in the appropriate container for disposal, ensuring that chlorinated solvents (including mixture of these with other solvents) are placed in the containers specifically identified for them. Spent alumina/silica should be placed in polyethylene bags and sealed.

Lachrymatory chemicals (e.g. benzoyl chloride, benzyl chloride and thionyl chloride) and substances with an objectionable odour must be used only in the fume cupboards.

Annexure - VI

Apparatus to be issued to each Student

Each student must have the following apparatus in a student's desk locker or drawer:

Two beakers-100 ml & 250 ml

Six test-tubes and six boiling tubes

Four Pasteur pipetts 8"

Four Pasteur pippets 6"

Six rubber bulbs

Four glass droppers

Two micro spatulas 6"

Two glass rods, 3 mm × 180 mm

One test-tube rack, two or three –row

One wire gauze

Sample bottles

Small scissors

One tripod stand

One forceps

One iron ring

Two small clamps

Three clamp holders

One clamp stand

One spirit lamp

One separating funnel 50 or 100ml

One test –tube brush

Two filter funnels 2”

One small box of filter papers

Two conical (Erlenmeyer) flasks 50 & 125 ml

Melting point capillaries 25 no.

One polyethylene wash bottle

Two watch glasses

One graduated cylinder 10 ml

Red and blue litmus paper books

Droppers

Glass tube 30 cm

Annexure - VII

Details of Elements

Element Name	Symbol	Atomic Number	Density	Atomic Mass
Aluminium	Al	13	2.70	26.98
Antomony	Sb	51	6.691	121.76
Argon	Ar	18	1.78	39.95
Arsenic	As	33	5.72	74.92
Barium	Ba	56	3.76	137.33
Beryllium	Be	4	1.85	9.01
Bismuth	Bi	83	9.8	209.98
Boron	B	5	2.35	10.81
Bromine	Br	35	3.12	79.90
Cadmium	Cd	48	8.65	112.41
Calcium	Ca	20	1.55	40.08
Carbon	C	6	2.2	12.01
Cerium	Ce	58	6.67	140.12
Chlorine	Cl	17	3.2	35.45
Chromium	Cr	24	7.19	51.99
Cobalt	Co	27	8.9	58.93
Copper	Cu	29	8.96	63.55
Erbium	Er	68	9.05	167.26
Europium	Eu	63	5.26	151.96
Fluorine	F	9	1.7	18.99
Germanium	Ge	32	5.32	72.64
Gold	Au	79	19.3	196.97
Helium	He	2	0.18	4.00
Hydrogen	H	1	0.09	1.01
Iodine	I	53	4.94	126.90
Iridium	Ir	77	22.7	192.22
Iron	Fe	26	7.86	55.85
Krypton	Kr	36	3.7	83.80

Table *contd...*

Element Name	Symbol	Atomic Number	Density	Atomic Mass
Lead	Pb	82	11.4	207.20
Lithium	Li	3	0.53	6.94
Lutetium	Lu	71	9.84	174.97
Magnesium	Mg	12	1.74	24.31
Manganese	Mn	25	7.43	54.94
Mercury	Hg	80	13.6	200.59
Molybdenum	Mo	42	10.2	95.94
Neon	Ne	10	0.9	20.18
Nickel	Ni	28	8.9	58.69
Niobium	Nb	41	8.57	92.92
Nitrogen	N	7	1.25	14.01
Osmium	Os	76	22.6	190.20
Oxygen	O	8	1,43	16.00
Palladium	Pd	46	12.0	106.42
Phosphorus	P	15	1.82	30.97
Platinum	Pt	78	21.4	195.08
Potassium	K	19	0.86	39.10
Protactinium	Pa	91	15.4	231.04
Radium	Ra	88	5.0	226.03
Rhodium	Rh	45	12.4	102.91
rubidium	Rb	37	1.53	85.47
Ruthinium	Ru	44	12.4	101.07
Scandium	Sc	21	3.0	44.96
Selenium	Se	34	4.79	78.96
Silicon	Si	14	2.33	28.09
Silver	Ag	47	10.5	107.87
Sodium	Na	11	0.97	22.99
Strontium	Sr	38	2.6	87.62
Sulphur	S	16	2.07	32.06
Tantalum	Ta	73	16.5	180.95
Tellurium	Te	52	6.24	127.60
Thallium	Tl	81	11.85	204.38
Thulium	Tm	69	9.33	168.93
Thorium	Th	90	11.7	232.04
Tin	Sn	50	7.30	118.69

Table *contd...*

Element Name	Symbol	Atomic Number	Density	Atomic Mass
Titanium	Ti	22	4.54	47.88
Tungsten	W	74	19.3	183.85
Uranium	U	92	19.07	238.03
Vanadium	V	23	6.1	50.94
Xenon	Xe	54	5.89	131.29
Yttrium	Y	39	4.47	88.91
Zinc	Zn	30	7.13	65.38
Zirconium	Zr	40	6.49	91.22

Annexure - VIII

Questions and Solutions

1. What is the best method to distinguish between the alkane and alkene?
2. Benzene C_6H_6 is the first member of homologous series with the general formula C_6H_{2n-x} (n is the number of carbon atoms in one molecule). What is the value of x?
3. Suggest why explorers take canisters of propane and not butane when they are on the expedition to the Antarctic?
4. Fractional distillation works in the case of hydrocarbons because:
5. Why are long-chain hydrocarbons better lubricants for car engines than short–chain hydrocarbons?
6. Explain how the fractionating column can separate the components of a mixture according to their different boiling points.
7. Why is most vegetable oils liquid at room temperature?
8. What is the bio-product on saponification of vegetable oils?
9. What is the function of an emulsifier?
10. Why most vegetable oils be hydrogenated to make margarine?
11. What do you understand by the term poly unsaturated?
12. Ethanol is oxidized to from substance X. X Reacts with ethanol to form Y. What are substances X and Y.
13. Which technique could reveal the relative molecular mass of a compound?
14. What is the difference between qualitative and quantitative analysis?
15. How is $R_f$value of a spot on a chromatogram calculated?
16. What spectral technique is used to identify the 'functional groups' in an organic compound.
17. When a carboxylic acid and an alcohol reacts in the presence of an acid catalyst, what is the product?
18. Give two uses of ethanoic acid.

19. (a) Give an example of a combined analytical technique that could be used to identify the substances present in the athlete's urine sample, if they are guilty of taking performance-enhancing drugs.

    (b) Explain how the different substances are separated using this technique.

20. The rate of reaction approximately doubles for every 10 ^{0}C rise in temperature. How much faster is a reaction at 100 ^{0}C than at 30 ^{0}C?

21. Explain the following observations

    (a) S orbitals are non-directional

    (b) CO_2 molecule has a zero dipole moment

    (c) Ammonia has much larger dipole moment than NF_3

    (d) Although ethanol and dimethyl ether are isomeric, they differ markedly in their boiling points.

22. Benzene contains three double bonds but it does not react with alkaline $KMnO_4$ and bromine in CCl_4. Explain.

23. Three compounds A, B, and C having the molecular formula C_8H_{10} theoretically can have one, two and three mono-bromo derivatives. What are A, B, C and their bromo derivatives?

24. Why methyl sulphate is a good methylating in the following reaction

    $$C_6H_5\,OH + (CH_3)_2SO_4 + aq.NaOH \rightarrow C_6H_5O\;CH_3 + CH_3OSO_3^-Na^+$$

    The good leaving group methyl sulphate anion ($CH_3\;SO_4^-$) in dimethyl sulphate makes it a good methylating agent.

25. Addition of bromine to *cis*-2-butene gives a d l pair whereas addition to *trans* -2-butene gives a *meso* compound. Explain?

26. Indicate whether the following compounds undergo hydrolysis by SN^1 or SN^2 with aqueous sodium hydroxide.

    (a) $CH_3\;Cl$ (b) $(CH_3)_3 - C - (Cl)$

    (c) $C_6H_5\;CH_2\;Cl$ (d) $CH_2 = CH - CH_2 - Cl$

27. What is red and blue shifts in UV spectrum?

28. Define what is meant by shielded and deshielded.

29. How would you differentiate the following two compounds by a chemical test?

30. Complete the following equations: Name each organic product.

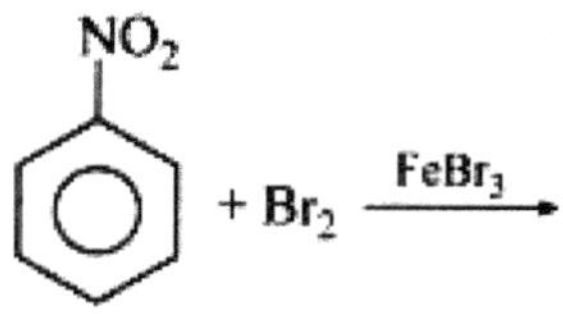

(a)

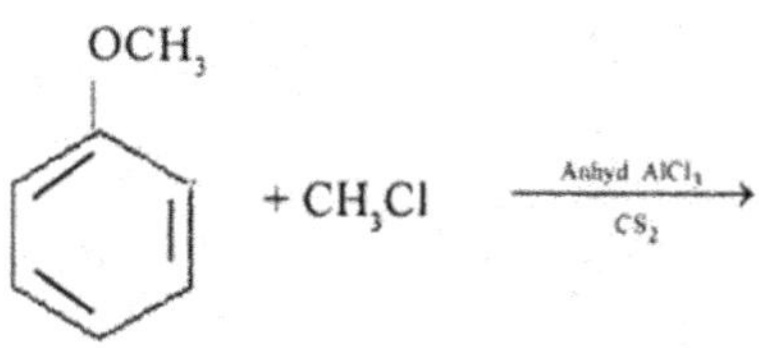

(b) Anisole

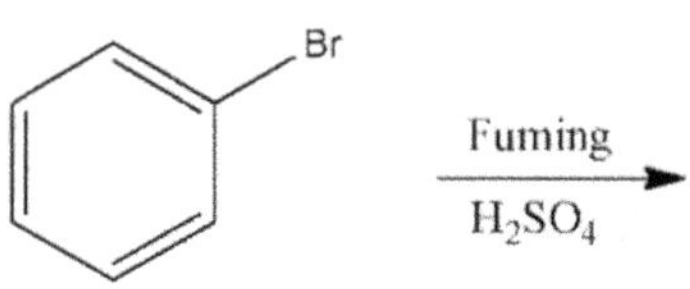

bromobenzene

(c)

31. How would you synthesize the following from benzene?
    (a) *para*-bromonitrobenzene
    (b) *meta*-bromonitrobenzene

Solutions:

1. Add bromine water, alkenes decolourise bromine water where as alkanes do not react.
2. Six (6)w
3. Butane has a higher melting point than propane and would turn into liquid.
4. Different hydrocarbons have different boiling points.

5. They are more viscous and have higher boiling points.
6. Vapours enter the column and travel upwards, from level to level until they reach the temperature at which they condense. Different fractions condense at different levels.
7. They have carbon-carbon double bonds. (C=C)
8. Glycerol
9. It helps an insoluble oil or fat mix fully with water forming an emulsion
10. The molecules have to lose some of C=C bonds, so that they can solidify at room temperature.
11. Molecules which have more than one C=C bonds.
12. X = CH_3COOH Y = $CH_3COOC_2H_5$
13. Mass spectrometry.
14. Qualitative analysis identifies what is present, quantitative measures how much.
15. Distance moved by the spot divided by distance travelled by the solution.
16. Infrared spectroscopy
17. An ester.
18. (i) Vinegar to preserve food or flavor food.

    (ii) Ethanoic acid is used to prepare esters.
19. (i) Gas chromatography and mass spectrometry.

    (ii) Different substances have different retention times as the sample passes through the column, depending on their attraction to the stationary phase.

    Modern analytical techniques are much more sensitive and can detect more substances and much lower concentration.

    Modern analytical techniques are much more sensitive and can detect more substances and much lower concentration.
20. 128 times faster.
21. (a) S orbital are non-directional, since they are spherically symmetrical.

    (b) The polarity of a molecule depends not only on the polarity of its individual bonds but also upon their direction i.e., upon the shape of the molecule. Here although the individual C-O bond is polar, the net polarity of the molecule will become zero due to its linear shape. So CO_2 has a zero dipole moment.

(c) In NH_3 as well as in NF_3, the unshared pair of electrons occupy the SP^3 orbital.

In case of NH_3 the contribution of the lone pair is not opposed to the other bonds, whereas in NF_3 the lone pair contributes a dipole moment in the direction opposite to that of the other three N-F bonds. These opposing moments are evidently of the same size and the result is a small dipole moment.

(d) Ethanol has intermolecular hydrogen-bonding thus decreasing its vapor pressure and hence its boiling point is high. In dimethyl ether, CH_3-O-CH_3, there are no such interactions and boiling point is low.

22. Benzene resists addition in which its ring system (aromatic stability) would be destroyed. Therefore, it does not undergo addition reactions with reagents like $KMnO_4$ and Br_2/CCl_4.

23. (i) 1,4-Dimethyl benzene (A)

Compound A gives only 2-bromo derivative

(ii) 1,2- Dimethyl benzene (B)

Compound B gives two bromo derivatives (3rd and 4th positions)

(iii) 1,3 – Dimethyl benzene (C)

Compound C gives two bromo derivatives (2nd and 4th positions)

24. $C_6H_5OH + aq.NaOH \rightarrow C_6H_5\,O^-Na^+ + CH_3\,OSO_2OCH_3 \rightarrow$

$C_6H_5OCH_3 + Na^+\,O^-SO_4.CH_3$

25. The addition of bromine is stereoselectively *trans* via bromonium ion. Since the bromonium ion is symmetric, the bromide ion can attack from the rear side, either carbon atom equally well, thereby resulting in the formation of d l pair.

In the case of *trans*-2-butene, the attack of the bromide ion on either carbon will lead to the same product which is a *meso* compound.

26. (a) S_N^1 (b) S_N^2 (c) S_N^1 (d) S_N^1

27. A shift of absorption to a longer wavelength is called a bathochromic shift (red shift). A shift of absorption to a shorter wavelength is called a hypsochromic shift (blue shift).

28. A molecule giving rise to a signal further up-field than another nucleus is set to be more shielded. The conclusion is that shielded protons absorb radiation at higher fields (frequency) while the deshielded protons will absorb at lower fields (frequency).

29. Treat with Tollen's reagent. The aldehyde will give a silver mirror test; the ketone will not.

30. (a)

(b)

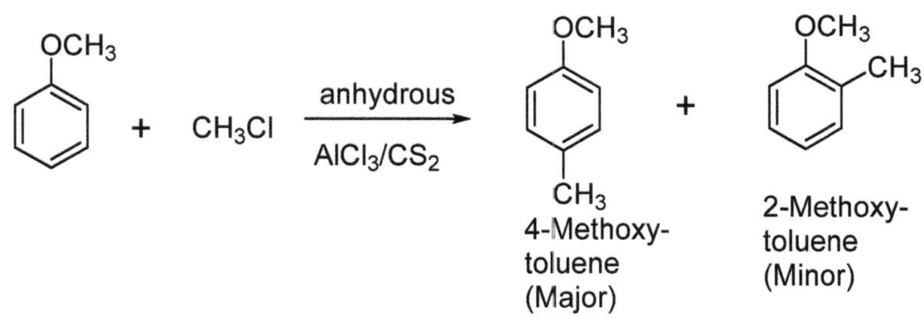

(c)

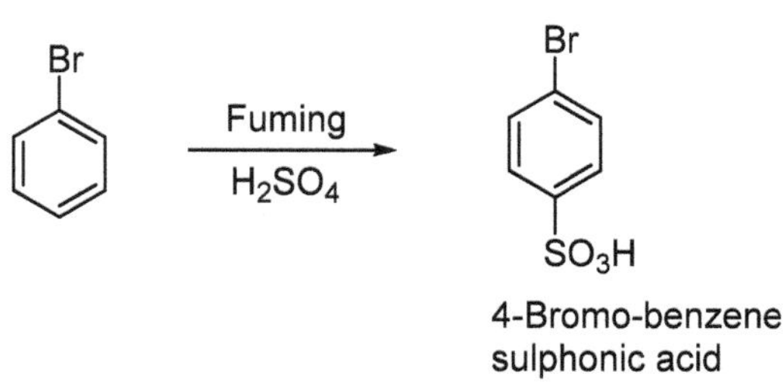

31.

(a) *para*-bromonitrobenzene

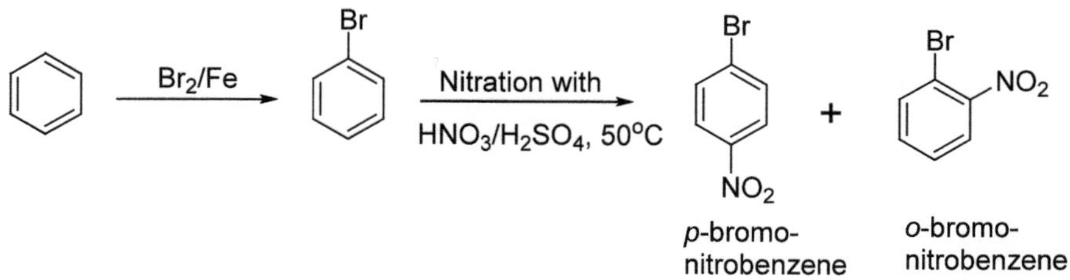

(b) *meta*-bromonitrobenzene

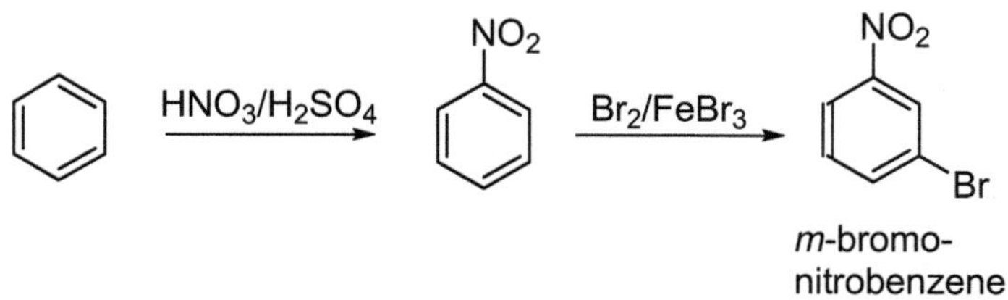

Bibliography

1. Organic experiments, L. F. Fieser, D. C. Heath & Company, Boston, 1941.
2. The systematic identification of organic compounds, R. L. Shriner; R.C. Fusion and D.Y. Curtin., John Wiley & Sons inc., New York, 1956.
3. Laboratory manual of organic chemistry, B. B. Dey and M.V. Sita Raman (Revised by T.R. Govindachari) 1957, S. Viswanathan, Madras.
4. Identification of organic compounds, N. D. Cheronis and J. B. Entrikin; Inter Science Publications, John Wiley & Sons, New York, 1963.
5. Stothers. J. B., Chap. 4 in Elucidation Structures by Physical and Chemical Methods, K. W. Bentley, Ed., Vol. XI, Part I, Inter Science, New York, 1963.
6. Brand J. C. D., and G. Eglinton, Application of Spectroscopy to Organic Chemistry, Oldbourne, London, 1965.
7. Spectrometric Identification of Organic Compounds by R. M. Silverstein and G. C. Bassler, by John Wiley & Sons Inc., 2nd., ed., New York, 1967.
8. Mathieson, D. W., Ed., Nuclear Magnetic Resonance for Organic Chemists, Academic Press, New York, 1967.
9. Laboratory Exercises Organic Chemistry, By James M. Sugihara: Burgess publishing company, 1969.
10. Unitized experiments in organic chemistry, R. Q. Brewester, C. A. Vander Werf and W.C. McEwen; D. Van Nostrand and Co. inc., Princeton, N. J., 1977.
11. Systematic Catalogue of NMR Spectra by A. Ault & M. R. Ault, published by Universal Science Books, 1980.
12. Laboratory Manual, (1985), Pure and Applied Chemistry, University of Strathclyde. Glasgow, Scotland.
13. A Text book of practical organic chemistry including Qualitative analysis. A. I. Vogel, Longmans, Green & Co LTD, London, 5th edition 1989.

14. Vogel's Text Book of Quantitative Chemical Analysis., G. H. Jeffery, J. Bassett, J. Mendham and R. C. Denney., published by Longman Scientific & Technical, 1989.

15. Spectroscopic Methods in Organic Chemistry by D.H. Williams & I. Fleming, published by McGraw-Hill Education, 1995.

16. Spectral Data for Structure Determination of Organic Compounds by E. Pretsch, T. Clerc, J. Seibel & W. Simon, published by Springer, 2nd Edition, 1998.

17. Foundation of spectroscopy by S. Duckett and B. Gilbert, OCP Publishers, 2000.

18. Practical Pharmaceutical Chemistry by A. H. Buckett & J.B. Stenlake, published by Bloomsbury Academic & Professional, 2001.

19. Organic Structures from Spectra by L.D. Field, S. Sternhill & j. R. Kalman, 4th Edition, JOHN WILEY AND SONS, LTD, 2008.

20. Hand book of organic analysis, H.T.C. Clarke, Edward Arnold (publishers) LTD, London, 2015.

21. Introduction to spectroscopy by D. L. Pavia et al., Published by Cengage Learning India, Pvt. Ltd., 2015.

Index

F

G

H

I

L

M

N

O

P

www.ingramcontent.com/pod-product-compliance
Ingram Content Group UK Ltd.
Pitfield, Milton Keynes, MK11 3LW, UK
UKHW021327190726
13845UKWH00004B/52